Neuma Rúbia Figueiredo Santana
Antenor de O. A. Netto
Inajá F. de Souza

Environmental hydrodynamics in the lower São Francisco and anthropic actions

Neuma Rúbia Figueiredo Santana
Antenor de O. A. Netto
Inajá F. de Souza

Environmental hydrodynamics in the lower São Francisco and anthropic actions

The mouth of the São Francisco River in the environmental context

ScienciaScripts

Imprint

Any brand names and product names mentioned in this book are subject to trademark, brand or patent protection and are trademarks or registered trademarks of their respective holders. The use of brand names, product names, common names, trade names, product descriptions etc. even without a particular marking in this work is in no way to be construed to mean that such names may be regarded as unrestricted in respect of trademark and brand protection legislation and could thus be used by anyone.

Cover image: www.ingimage.com

This book is a translation from the original published under ISBN 978-613-9-65685-1.

Publisher:
Sciencia Scripts
is a trademark of
Dodo Books Indian Ocean Ltd. and OmniScriptum S.R.L publishing group

120 High Road, East Finchley, London, N2 9ED, United Kingdom
Str. Armeneasca 28/1, office 1, Chisinau MD-2012, Republic of Moldova, Europe
Printed at: see last page
ISBN: 978-620-7-79495-9

SUMMARY

AGRADEDIMENTO 2

SUMMARY 4

GENERAL INTRODUCTION 5

OBJECTIVES 6

CHAPTER 1: THEORETICAL BACKGROUND 7

CHAPTER 2: CHARACTERIZATION OF WATER QUALITY AT THE MOUTH OF THE SÂO FRANCISCO RIVER 25

CHAPTER 3: HYDRODYNAMICS OF THE SÃO FRANCISCO RIVER FLOOD WITH THE USE OF COMPUTATIONAL MODELING 67

CHAPTER 4: SALINITY AT THE MOUTH OF THE FRANCISCO RIVER USING COMPUTER MODELING 82

CHAPTER 5. ENVIRONMENTAL EDUCATION ACTIVITIES IN THE REGION OF THE MOUTH OF THE RIVER SAN FRANCISCO: SOCIOECONOMIC PROFILE OF FISHERS AND DEGRADATION OF THE ENVIRONMENT 101

APPENDICES 113

ANNEX..116

"I dedicate this work to my parents, siblings, nephews, uncles and grandparents. Especially to my grandmother Elvira Santana (In memoria)".

AGRADEDIMENTO

To my God who has been my teacher, father and companion during this period of knowledge. Thank you for your goodness, providence, support and strength to finish this research "To you my eternal gratitude and adoration".

In this work, helping hands have been present. That's why I would like to thank my parents, siblings and family for their encouragement, prayers, affection and encouragement during my participation in the courses, field trips and writing. Without you, I believe it would have been much more difficult to complete this academic stage.

To my advisor Professor Dr. Antenor de Oliveira Netto for taking on this challenge and for his constant encouragement of scientific and professional knowledge.

To the fellows Samuel Barreto and José Carlos Benicio for their extreme cooperation in the collections and in the daily routine of the research: Congratulations on your work.

To the coordinator of Prodema, Prof. Dr. Maria José Soares Nascimento, and to the professors of the center and the secretariat.

To Professors Inaja Francisco de Souza and Carlos Alexandre Borges Garcia for their enormous contributions to the development of this work.

To the researchers of the Acqua/UFS research group Anderson Nascimento do Vasco, Marinoé Gonzaga da Silva, Gregorio Guirado Faccioli, Thassio Monteiro M. da Silva and especially Edson Leal Menezes Neto, Ricardo Castillo Salazar, Carlos Prata, Fabio Brandâo Brito, Sérgio Araùjo, Jorge Yonuma Hotel Bonga, Kenneth Michael Doll, Joâo Marcos and Crystiano Ayres.

The Aguas do Sâo Francisco project team - Thadeu Ismerim Silva Santos, Ricardo Rogério da Silva Santos, Èrica de Oliveira and Flavia Nunes.

To the driver, Mr. Amaral, for the smooth and provincial journeys.

To my friends from near Aracaju/SE and far away Salvador/BA, I found support and laughter in you.

Professor Carmem Regina Guimarâes of the UFS Biology Department for her help with the field equipment.

To the Georiomar/UFS research group, especially Professor Luiz Fontes and Jonas Ricardo dos Santos.

To Professor Carlos Alexandre Borges Garcia, through the Environmental Chemistry Laboratory, and to Ana Carla, the technician, for their attention and support during collection and analysis.

To Professor Yvonildes Medeiros for allowing us to use the UFBA Water Resources Department to carry out the SisBahia training. To master's student Tiago Rosario for his training and assistance in running the hydrodynamic model.

Sergipe's water resources secretariat

FAPITEC and CAPES for the doctoral scholarship and financial support during the research work.

Finally, to everyone who directly or indirectly made this work possible. Thank you

SUMMARY

The hydrodynamics of the waters of a river flowing to the sea establish an important control on the volume and direction of the watercourse, among them salinity, a physical-chemical property that represents the measure of the concentration of dissolved salts in the body of water. This is because water reflects the environmental conditions of a river. The aim of this study was to assess the hydrodynamics of the estuary and their relationships. The mouth of the Sâo Francisco River is located between the states of Sergipe and Alagoas. It is a fluvial-marine plain made up of varzeas and river and marine terraces, represents an area of extreme natural beauty and economic importance for the local population and is part of the geographical division of the lower Sâo Francisco. Using a boat located in the municipality of Brejo Grande (SE) and with the help of a Garmin e-trex digital GPS, latitude and longitude information was recorded in UTM to map the water collection points. The SiSbahia program was used for environmental modeling of salinity and flow coefficients. In order to check water quality and degradation processes, the sampling points were identified by measuring them with a HI9828 multi-parameter probe during the months of February, March, April, May, September, November 2015 and March 2017. The probe recorded salinity, electrical conductivity, turbidity, dissolved oxygen, total solids, temperature and pH. For drinking water, water was collected from households to check pH, sodium and chlorides. Interviews were conducted with residents of the Brejo Grande/SE and Saramém regions to ascertain their perception of salinity. The quantitative and qualitative data was then processed using statistical software and compared with CONAMA Resolution 357/2005 and Ordinance MS 2914/2011. The results presented for the environmental modeling showed the action of the tide after the municipality of Penedo/AL, the salinity recorded by the presence of the saline wedge was perceived 11 km from the mouth. The pH ranged from 6.80 to 9.47, with the highest value occurring in February 2015. Dissolved oxygen values in the area evaluated ranged from 4.68 to 9.27 ppm. Total dissolved solids values ranged from 7.0 mg.L^{-1} to 7522 mg.L^{-1} . The highest concentration of total dissolved solids recorded occurred at Point 1 in November, during a dry period with a sizig tide regime and a water level rise of 2.2m. The turbidity values recorded ranged from 2.9 NFU to 47.07 NFU. Electrical conductivity values ranged from 10.00 (µS/cm^{-1}) to 13130.00 (µS/cm^{-1}). The average concentration of salinity ranged from 0.17 to 28.87 (‰), with the highest values recorded on the shore of the municipality of Brejo Grande/SE and Points 39, 40 and 43 located towards Piaçabuçu in the state of Alagoas. The pH in drinking water between the municipalities of Brejo Grande/SE and the village of Saramém showed average values within the limit established by Ministry of Health Ordinance 2914/2011. In the evaluation of the sodium and chloride parameters, the average values obtained in May 2015 were between 1962.85 mg.L^{-1} and 2938.62 mg.L^{-1} . In response to the semi-structured interviews, residents reported that the change in the taste of the water was perceived during high tide. It can be concluded that the salinity concentrations were significant and indicated that the tide period was responsible for the increase in salinity. Salinity is a worrying factor because it directly compromises water use and affects the region's biota. The environmental education activity enabled the fishermen to understand the main types of environmental degradation in the region under investigation.

Key words: Environmental hydrodynamics. Salinization. Anthropic action. Environmental education.

GENERAL INTRODUCTION

Rivers are fundamental to the population and the balance of natural resources, but currently a large part of their tributaries are suffering the strong effects of anthropogenic actions, which signal the advance of environmental degradation in these ecosystems. These changes jeopardize the sustainability of large rivers, including the São Francisco river basin, which currently has many of its tributaries at risk. The São Francisco River went through a period of dam construction, which was responsible for natural and social changes in this area, reflecting the lack of in-depth studies of the environment during the installation period.

For Medeiros et al (2007), one of the most notable changes made by the construction of dams on rivers is the regularization of the flow, with the aim of supplying the water needed to generate hydroelectric power, causing a large reduction in the natural flow, causing an imbalance of energy between the river and the sea. Consequently, these changes affect the distribution of water for consumption and modify the aquatic ecosystem, through the recharge of sediment, nutrients and salinity.

According to Souza (2006), salinity is an important indicator of water quality and increased saline intrusion into a river can have adverse effects on the aquatic environmental system. In addition, salinity is one of the variables that guides the possible uses of water resources in a river basin, making it a challenge for managers (SANTOS, 2014). Oliveira et al. (2008) also point out that salinization can cause damage to the region's domestic supply and agriculture.

According to Aguair Netto et al (2011), these impacts can also compromise the economy and generate social problems for the riverside population and urban centers. Ferreira et al (2011) highlights the social and economic relevance of the lower reaches of the River San Francisco, which meets a demand for uses such as supplying riverside populations, irrigation, aquaculture, ecotourism and navigation.

In this way, water is a precious resource and must be maintained in conditions suitable for the use and maintenance of all forms of life on earth. Ensuring the quality of rivers is the challenge of the current century. Based on these points, the monitoring of physical, chemical and biological parameters is essential to the basin management process. This information is fundamental for drawing up plans and policies that can guarantee the supply of water, adequately meeting the multiple uses (AGUIAR NETTO et al. 2011).

OBJECTIVES

GENERAL: To characterize the environmental hydrodynamics and anthropogenic actions at the mouth of the São Francisco River.

SPECIFIC:

S Carry out qualitative water monitoring in areas of the mouth of the River San Francisco.

S To model and simulate the behavior of hydrodynamics with the SisBahia computer model at the mouth of the São Francisco River and to develop scenarios of the velocities of the flood and ebb tides in the region under study.

S Model and simulate the behavior of salinity with the SisBahia computer model at the mouth of the São Francisco River and develop scenarios for the distribution of salinity in the region under study.

S To verify the social and environmental impacts of environmental education activities in areas of the mouth of the River San Francisco.

ORGANIZATION OF THE THESIS

The thesis will be structured in chapters, which will be organized as follows:

In chapter 01, the theoretical background will be discussed with the main themes: Sustainable development; Hydrographic basins *of* the San Francisco River; Environmental modeling and hydrodynamics; and Water quality.

S In chapter 02 Assessment of water quality at the mouth of the River San Francisco

S In Chapter 03 Salinity at the mouth of the São Francisco River using computer modeling.

S In Chapter 04 Hydrodynamics at the mouth of the San Francisco River using computer modeling

S In Chapter 05 Environmental education activities in the region of the mouth of the River Sao Francisco
Francisco: socio-economic profile of fishermen and environmental degradation

Finally, the final considerations, bibliographical references, Appendices and Annexes.

CHAPTER 1: THEORETICAL BACKGROUND

1.1 Sustainable development and water resources

Sustainable development is the basis for the use of any natural resource and necessarily the water resource, as it is responsible for sustaining life on the planet and is one of the means for the satisfactory performance of urban and rural development. Silva and Mendes (2005) state that the concept of sustainable development is linked to an increased concern with the maintenance and existence of natural resources and a suitable environment for the continuity of future generations.

Conceptually, the intention is for sustainable development to establish threads in the main processes necessary for human activity, but without forgetting the conservation and limits on the use of natural resources. For Setti and Bógus (2010), this relationship points to a concern in the relationship between man and nature in recent decades, raising debates in the context of evaluations of the perverse effects of the economic development model throughout the world.

These debates gained momentum after the Stockholm conference in 1972. Based on these dialogues, the United Nations Conference on the Environment considered the need for a common vision and principle to guide humanity in conserving and improving the human environment (UNEP/United Nations Environment Programme, 2016). This argument was consolidated by proclaiming in a declaration that man is both the work of and the builder of the environment in which he lives, which provides him with material sustenance and at the same time offers him the opportunity to develop intellectually, morally, socially and spiritually (UNEP, 2016).

Lago (2006) highlights this event as a liberation movement to free man from the threat of his enslavement to the dangers he himself created for the environment in which he lives. The author also stresses that this moment was a historic stage in the evolution of the treatment of environmental issues at international level. However, as the issue gained increasing international legitimacy, it began to be discussed less and less from a scientific point of view, and more and more in a political and economic context (LAGO, 2006).

From the perspective of global sustainable development, in 1987 the World Commission on the Environment (WCED) presented a report entitled Our Common Future, which represented one of the first global efforts to compose a global agenda for a paradigm shift in the human development model (MADEIRA, 2014).

This report points out that concern for the environment can be conducted in cooperation between developing countries, and between countries at different stages of economic and social development, leading to the achievement of common goals and mutual support that take into account the interrelationships between people, resources, the environment and development (BRUNDTLAND, 1987).

Gallo and Setti (2012) report that sustainability provides guidelines both for a rationality that guarantees solidarity and global cooperation, and for the continuity of development and life itself for future generations. Setti and Bógus (2010) also explain that the environmental, social and economic problems to be tackled include the idea of "sustainable development" as a strategy for reversing the world's poverty, deforestation and social inequality.

In this sense, foundations are being laid for the construction of a global society, in which fair equity and a clear concept of environmental development can be established. The first milestone is the document entitled *"The Earth Charter", a* commitment devised at the 1992 United Nations Conference on Environment and Development. This document was drafted in the 1990s, through open and participatory consultation, affirming a pedagogy of sustainability, where it recognizes that environmental protection, human development, human rights and peace are interconnected components for harmony between society and the environment (SILVA, 2011).

Rattner (1999) states that one of the most noticeable results of international conferences in recent decades has been the incorporation of sustainability into development debates. However, the concept of sustainability transcends the analytical exercise of explaining reality and requires the testing of logical coherence in practical applications. In order to disincorporate the formula currently used in political and scientific discourse, "economically viable, socially equitable and ecologically sustainable", since it does not take into account the ways and means of combining and integrating goals on technical progress and productivity with the protection and conservation of natural resources and the environment (RATTNER, 1999).

The use of interdisciplinarity is an extremely complex context in which to apply sustainability. Foladori (1999) describes that economic, environmental and social variables are applied in their own methodologies. In other words, the concept of sustainable development cannot only provide future generations with a material world (biotic and abiotic) that is equal in the present and the future, but must also relate to equity in social relations, making interconnection possible.

This interconnection in origin means a paradigm shift where the effort is represented in the unity of man and nature. Sparemberger and Marques (2015) look at development planning and believe that it is essential for social sustainability to build a society of 'being', in which there is a distribution of 'having', in order to improve the conditions and rights of the population.

Nascimento (2012) states that a sustainable society means that all citizens have the minimum necessary for a dignified life and that no one absorbs goods, natural and energy resources that are harmful to others. Therefore, referring these principles to water resource management, water and related resources should be managed for human well-being, contributing to poverty reduction and economic growth (WWDR, 2015).

Castro and Scariot (2008) point out that access to water is already one of the limiting factors for socio-economic development in many regions. They also state that the absence or contamination of water bodies leads to a reduction in living spaces and, in addition to immense human costs, causes a global loss of social productivity, leading to competition for its various uses.

Currently in Brazil, water resources are managed by hydrographic basins throughout the country, either in bodies of water owned by the Union or by the states (PORTO and PORTO, 2008). Hydrographic basins, according to Santana (2003), refer to a natural geographical compartment delimited by water dividers. This compartment is superficially drained by a main watercourse and its tributaries. In general, a hydrographic basin can be defined as a natural geographical area, delimited by the highest points of the relief, within (among) which rainwater is drained superficially by a main watercourse until it leaves the basin at the lowest point of the relief, which corresponds to the mouth of that watercourse.

From a hydrological point of view, these concepts provide a definition of a hydrographical basin, based on the natural phenomena found in the hydrological cycle, such as precipitation, evaporation, infiltration and runoff. However, the hydrographic basin should not only be understood as a recipient and disperser of water (TUCCI, 2005), but also the current conditions of water availability and demand for this natural resource. Observing climatic conditions, land use, preponderant uses and their priorities, as well as the establishment of monitoring points for the quality and quantity of water for certain purposes and the maintenance of environments.

According to Tucci, Hespanhol and Cordeiro Netto (2000), the Brazilian Northeast has unfavorable hydric conditions that combine: high evapotranspiration throughout the year, low rainfall, unfavorable subsoil in many regions (brackish water or crystalline formation) and low social and economic development.

The region of the lower Sao Francisco River is part of the Brazilian Northeast, and is currently facing major environmental problems, which according to Ferreira et al. (2011), are significant changes in the dynamics of the river, loss of fauna and flora, as well as erosive processes on its banks promoting siltation in its bed.

Santos (2009) also highlights other socio-environmental impacts that have occurred in the lower reaches of the River San Francisco, such as the extinction of the natural flood cycle of the River San Francisco itself, the suppression of the exuberant varzea hydroforest and the lowering of the water table that existed in the area during the natural flow of the river, and due to the lack of this flow, the extinction of the rice cultivation cycle on which the community depended for its survival.

Another significant change was the decline of artisanal fishing in favor of industrial fishing. According to Pacheco and Lira (2004), the practice of artisanal fishing was widely used in the flood season when the banks were invaded by the waters, feeding the varzeas and small marine lagoons with organic matter and various species of fish and shrimp, which were raised extensively with complementary feed, made from agricultural by-products, mainly derived from rice. However, the author adds that the action of damming various stretches of the river has altered the periodic flooding which has been replaced by water flow control, affecting fish reproduction in the region.

These scenarios jeopardize the sustainability of this region's water resources, as well as the socio-economic quality of the population. Given that factors such as the salinity of its waters, the discharge of nutrients from industrial aquaculture waste, the lack of basic sanitation and erosion along the riverbank, drastically affect the ecological balance and environmental health of the localities at the mouth of the River San Francisco.

1.2 São Francisco river basin

Tucci (2007) considers the hydrographic basin to be a physical system in which the input is the volume of water precipitated, and the output is the volume of water drained by the outlet, taking into account intermediate losses, evaporated and transpired volumes and also deep infiltration.

However, Santana (2003) reports that the term hydrographic basin refers to a natural geographical compartment delimited by water dividers. This compartment is superficially drained by a main watercourse and its tributaries. Generally speaking, a hydrographic basin can be defined as a natural geographical area, delimited by the highest points of the relief, within which rainwater is drained superficially by a main watercourse until it leaves the basin, at the lowest point of the relief, which corresponds to the mouth of that watercourse.

Araùjo (2010) stresses that research that analyzes hydrographic basins as a study unit is essential, since it is a well-characterized physical unit, both from the point of view of integration and the functionality of its components, and that the analysis of the environment based on the study of hydrographic basins is a strong tool for the development of a concept in the treatment of socio-environmental issues, allowing the researcher to verify the possible alterations caused by communities.

In this sense, Cardoso et al. (2006) point out that in order to assess the effect that any disturbance would have on a hydrographic basin, it is first necessary to know the hydrological characteristics of the natural ecosystem very well, in order to later compare them with those in which there has been anthropogenic intervention. For this reason, Botelho and Silva (2004) summarize that the hydrographic basin could be considered a privileged space where the most important environmental interactions occur, through the interaction of water with other natural elements, and with anthropic activities, and can be pointed out as the most appropriate territorial unit for the management not only of water resources, but of an integrated environmental management whose ultimate goal is sustainable practices.

Nascimento and Vilaça (2008) also stress that hydrographic basins have been adopted as physical units for recognition, characterization and evaluation, in order to facilitate the approach to water resources. The behavior of a hydrographic basin over time is due to two factors: natural, responsible for the environment's predisposition to environmental degradation, and anthropic, where human activities interfere directly or indirectly in the functioning of the basin.

For this reason, river basin planning has developed in response to the new demands made by society or economic growth, which require quality water for its various uses. In the historical context, river basins are considered privileged areas for promoting regional development, as they have an intense diversity of renewable and non-renewable natural resources (ANA, 2014).

The São Francisco river basin has these characteristics, as well as being an integral part of the country, offering a wide range of natural goods and services, and if it is treated sustainably, it contributes to the country's economic and ecosystem balance (ANA, 2014).

With the arrival of the first Europeans in Brazil, the São Francisco River provided the easiest and fastest means of contact between the colonies. This was because the Portuguese colonies were established in separate points, and contact by land was very difficult, due to the density of the forest and the presence of hostile indigenous people, making it possible to take risks during the journey (PIERSON, 1972). Soon afterwards, this river, which bathes the states of Alagoas, Bahia, Pernambuco, Sergipe, Minas Gerais, Goias and the Federal District, was considered the river of 'national integration', leaving a legacy of the importance of this water resource for the Brazilian people.

According to Ramina (2014), the hydroelectric use of the San Francisco River began around the 1950s and 1960s, in a demographic, economic, political and institutional environment that was very different from the current one. Gonçalves and Oliveira (2009) argue that a large part of the use of the waters of the San Francisco River is associated both with the need to increase energy capacity, through hydroelectricity with the construction of large reservoirs, and with large irrigation projects.

Although the generation of electricity and irrigation are of great economic importance to the country, it is worth noting that the flow of the River San Francisco is decreasing, and several socio-environmental problems are already emerging in all its dimensions, but today they are most evident in the area of the mouth between the states of Alagoas and Sergipe. Medeiros et al. (2011) point out that hydroelectric exploitation is among the main causes of impacts on the São Francisco river basin.

In all, there are nine hydroelectric plants along the São Francisco River, responsible for visible and drastic changes to the river's ecosystem, causing immense social impact, altering the flood and ebb regimes and compromising the reproduction of species (ZELLHUBER and SIQUEIRA, 2007). The aforementioned authors add that these changes began to occur because the flow of water was established by hydroelectric power stations, completely altering their flooding regimes during the rainy season.

The existence of these problems causes anthropogenic pressure, generating a continuous process of degradation in the basin. It is known that the long-term average specific flow decreases from the source to the mouth of the river, which further affects the availability of water to meet the uses of the São Francisco river basin (KOSMINSKY and ZUFFO, 2009).

It should be noted that the ten-year water resources plan for the São Francisco River basin (PHBSF) indicates that the maintenance of the aquatic ecosystem requires flows of no less than 30% of the average flow, but irrigation accounts for 86% of the flow consumed and some conflicts between water uses have been established (KOSMINSKY and ZUFFO, 2009).

In order to mediate these conflicts, Martins et al. (2011) suggest the use of water availability analysis over time, in rivers that show changes in the flow regime as a result of anthropogenic interventions, as is the case with the São Francisco River, making it essential to guarantee the long-term exploitation of the resource. It is worth noting that initially the main use of the Sao Francisco River was navigation, which served as a means of communication for the flow of regional production, but it is essential to pay attention to other uses such as: human supply, agriculture, fishing and livestock, (RAMINA, 2014).

From the point of view of extension and diversity, the Sao Francisco River area is divided into four physiographic units: the Upper Sao Francisco, which covers the source up to the city of Pirapora (MG); the Middle Sao Francisco, which extends from the city of Pirapora to Remanso (BA); the Middle Sao Francisco, which runs from Remanso to the city of Paulo Afonso (BA); and finally, the Lower Sao Francisco, which is delimited between Paulo Afonso and the mouth (PEREIRA et al. 2007).

Although electricity generation and irrigation are of great economic importance to the country, it should be noted that the flow of the San Francisco River is decreasing, and several socio-environmental problems are already emerging. These problems are the result of a lack of careful analysis of possible ecological impacts, coupled with a program of agricultural modernization with dominant economic aspects, giving little importance to environmental and social impacts (ANDRADE, 1984).

1.3 Environmental modeling and hydrodynamics

A model is something that can be reproduced, it is a small-scale representation of a current environment which includes future projections. The need for scenario simulations began as environmental degradation threatened the quality and availability of natural resources.

Environmental degradation threatens the global balance of natural compartments, so mathematical and computational forecasting of these phenomena can generate models that can be used in environmental monitoring and management programs.

Models are integrative tools that provide a dynamic view of processes in complex environmental systems (SAMPAIO 2010).

For Santos (2007), the complexities of environmental systems are linked to the large number of interactions between physical, chemical and biological phenomena and the mathematical equations that govern them. In general, these are not linear and require powerful software to model them properly, which is why the use of models has spread significantly in recent decades.

The advantages of using modeling in terms of research can be seen in the advances made in various areas of knowledge, including physics, chemistry and biology, among others, presupposing multidisciplinarity (BASSANEZI, 2002). Bassanezi (2002) carefully warns that no model should be considered finite, as it can be improved and that a good model is one that allows new models to be formulated.

According to Christofoletti (1999), modeling is a scientific methodological tool which consists of constructing a set of techniques in order to draw up a simplified picture of man's reaction to the environment around him.

According to Felgueiras (2001), the modeling process considers that natural processes are the result of complex spatio-temporal interactions between the various elements that make them up. Thus, in a mathematical model of a process, environmental properties are treated as variables in the model, while their interrelationships are represented by arithmetic or logical operations.

These models can be used to spatially and temporally interpolate and extrapolate the information obtained at a few points where the data was measured to the entire area of interest. And once the models have been calibrated, they are able to reproduce values at the points where the measurements took place, so that they can be used to interpolate and extrapolate information beyond those points, by simulating scenarios (ROSMAN, 2014).

Amorin and Silva (2009) simplify the concept of environmental modeling as the process by which the researcher transforms part of the information from the real world into the computational world. Thus, modeling applied to water resources can involve both the transport of substances and the hydrodynamics of the water body itself (SOUZA et al. 2011).

The first simulations that gave rise to environmental modeling occurred through mathematical modeling carried out by researchers Street and Phelps in 1925, who proposed the objective of simulating the oxygen profile in rivers and estuaries (BATISTA, 2016). To this day, this model provides the basis for the development of new research such as that by Bezerra, Mendonça and Frischkorn (2008); Haider, Ali and Haydar (2012); Ricciardone, Pereira and Pereira (2011) and Kannel et al. (2011).

As for hydrodynamic models, these came into force in the 1960s, with studies carried out on the propagation of waves that were influenced by nuclear explosions and their effects on coastal zones in areas of the Rhine estuary in the North Sea of Europe (LEENDERTSE, 1967).

Over the years, various works on hydrodynamic modeling have emerged and improved with the advance of computer technology, including Al-Rabeb, Gunay and Cekirge (1990), who simulated wind-driven seas in the Persian Gulf using HYDRO1 and HYDRO2.

As part of the process of improving the models, new tools have been added to the hydrodynamic evaluations, such as ecological and water quality simulations, which can be cited in various studies, including Cunha, Rosman and Monteiro (2003); Lillebo et al. (2005); Cunha, Ferreira,

Rosman and Teòfilo (2006); Ji et al. (2007); Lacroix et al. (2007); Liu et al. (2008); Shi and Lu (2011); Dutta et al. (2013); Sokolovaa et al. (2013); Zhon et al. (2015); Baird et al. (2016) and Ganju et al. (2016).

In Brazil, work on environmental modeling gained momentum after the publication of the national water resources policy. For Veiga and Magrini (2011), the promulgation of Law 9433/97, which instituted the national water resources policy, introduced important changes in the management of environmental quality and resources. Larentis (2004) adds that the use of modeling in conjunction with monitoring can help in the planning of water resources in a river basin, providing flexibility and reducing costs.

Modeling has been one of the necessary tools for managing water resources, and Brazil has sought to develop and incorporate these attributes to prevent and visualize impacts on aquatic environments. Among the hydrodynamic models used in Brazil is HEC-HAS, software developed by the Hydrological Engineering Center (CEIWR-HEC) in the United States. The system contains three elements for one-dimensional 1D river analysis, including flow, sediment and water quality analysis (USACE, 2010).

This model was used by Souza, Collischonn and Tucci (2007) where they simulated the propagation of flood waves along a stretch of the São Francisco River; Paiva, Collischonn and Bravo (2011) in a network of channels simulating stretches of river; in the Una River in Pernambuco by Ribeiro et al. (2014); Almeida et al. (2016) in experimental channels at the Federal University of Goias.

Rosario (2014) also describes around 12 types of hydrodynamic models used in Brazilian research, including SOBEK, DIVAST, IPH-ECO, EFDC, DELFT-3D, ECOMSED, ELCOM, MOHID, POM, ROMS, MIKE and SisBahia, which can simulate coastal zones, reservoirs and rivers.

It should be noted that the tendency to choose a model must be justified by the object of study and the availability of data, which are essential factors for efficient modeling. For these reasons, among the models mentioned above, the SisBahia hydrodynamic model is seen as efficient for the reentrant and estuary areas, as it does not require a high level of computational capacity and has been fully developed under Brazilian environmental conditions and is in the public domain.

The SisBahia Environmental Hydrodynamics Base System was created in Brazil and has been used since 1987 in the Coastal and Oceanographic Engineering area of the Ocean Engineering Program, and in the Database Area of the Systems and Computer Engineering Program, both at COPPE/UFRJ (ROSMAN et al. 2001).

The SisBahia model features a FIST (Filtered in space and time) hydrodynamic model, which represents a modeling of free-surface water bodies composed of a series of hydrodynamic models (ROSMAN et al. 2001). The model also features the Lagrangian and Eulerian systems.

The Lagrangian model is based on measuring or determining the acceleration of a non-fixed point in space that follows the flow or trajectory of the movement, while the Eulerian model consists of observing a fixed point (ASSAD et al. 2009).

For Rosman (2011) the Eulerian Model of Advective-Diffusive Transport (METAD) integrated vertically (2DH), for passive and non-conservative scalars. Non-conservative scalars, which represent the majority of substances in water, undergo concentration changes through physical, biological and chemical processes.

The transport processes, advective and diffusive, are solved in the same way as METAD. Biological and chemical processes, called kinetic reactions, are defined for each substance and are therefore the distinguishing feature of this model. The understanding and formulation of these processes are fundamental to the construction of the water quality model (ROSMAN, 2011).

As the hydrodynamics that occur in estuary areas are dynamic and driven by tidal flows, the equations are non-linear and therefore more complex. The 3D version of FIST solves the complete Navier-Stokes equations[2] , with shallow water approximation and uses an efficient numerical technique in two modules, first calculating the values of the free surface elevation

through vertically integrated two-dimensional modeling (2DH) and then the velocity field (ROSMAN, 2013).

FIST3D makes it possible to simulate hydrodynamic circulation in natural bodies of water under different meteorological, oceanographic, fluvial or lacustrine scenarios (XAVIER, 2002).

The model also features Water Quality Models: This is a set of Eulerian transport models for coupled simulation of up to 11 water quality parameters,

The SisBahia model has been used in the Bay of Todos os Santos by Xavier (2002); in estuarine areas in the Bay of Vitória do Espirito Santo by Rigo (2004); on the coast of the metropolitan region of Recife/PE by Rolnnic (2008); Fernandes (2010) at the mouth of the Amazon River; in coastal waters of Fortaleza Pereira (2012); estuarine system of the Tinharé-Boipeba archipelago/BA by Barboza et al (2014) and in the evaluation of salinity in the Almada River Itabuna/BA Santos (2014).

1.4 Water quality

The degradation of water quality in Brazilian rivers has become a concern in the management of water resources. This is because, according to Souza and Gastaldini (2014), water quality reflects the environmental conditions of the hydrographic basin, so knowing the characteristics of water quality broadens the ecological knowledge of the ecosystem and makes it possible to detect alterations caused by human activity.

There are various processes that degrade the quality of a river's water, including the increase in

---

1 Fluid motion equation

sediment deposited in riverbeds, which causes a decrease in runoff and constitutes barriers to the penetration of light into these environments (DERISIO, 2007).

Another important parameter for checking the aquatic dynamics of rivers is the salinity content, since excessive values can compromise the reproductive process of aquatic species and make human consumption unfeasible.

Medeiros et al. (20013) add that salinity is a determining factor for the characterization and maintenance of the environment's productive processes, while the change in freshwater flow strongly affects the concentration of nutrients in general and all water quality parameters. In addition, the quality of the water downstream is influenced by the intrusion of salt water, especially by altering its salinity.

According to Branco (1993), anthropogenic action is the greatest aggression against nature, and these problems always involve water sources, compromising the current or future quality of the water that supplies the city. Lima (2001) states that the varieties of elements discharged into water bodies can be grouped into two classes: point source and diffuse.

Domestic and industrial waste make up the point source group because they are restricted to a single point of discharge, facilitating the collection system through channels or networks. In general, the point source of pollution can be reduced through appropriate treatment for subsequent discharge. Diffuse pollution is characterized by multiple discharge points resulting from runoff in urban and/or agricultural areas and occurs during rainy periods, reaching very high concentrations of pollutants.

According to Rebouças (2002), rivers are the available sources of water, from which the population can be supplied with their needs, but urban and rural development can compromise the quality of these waters by contaminating the surface runoff network with local sewage discharges, making the source unviable, requiring new water collection projects in more distant areas or the use of more intense water and sewage treatment techniques.

The waste present in domestic sewage is carried into rivers, making them sources of pathogenic micro-organisms. In order to assess these micro-organisms in the aquatic environment, the use of biological parameters is important for checking for pathogens. Much of our concern about the purity of water has been related to the transmission of diseases. There are various criteria for an indicator organism, one of the most important being that the organism is present in significant numbers in human feces, so that its detection is a good indication of human waste being introduced into the water (TORTORA; FUNKE and CASE, 2005).

Currently, the determination of coliform bacteria is one of the main indicators of biological contamination, which is why researchers have shown enormous interest in the use of biological evaluations, due to the significant response to possible natural and man-made interventions that may be occurring in a given natural resource.

According to Paula et al. (2013), in aquatic environments there are multiple sources of pollution and their dynamism over time and space involves a series of consequences. The different uses to which they are put require different qualitative characteristics.

Purity requirements vary according to the intended use of the water and bacteriological water quality standards are based on consumer protection in order to avoid waterborne diseases, most of which are caused by pathogenic microorganisms of enteric origin via the fecal-oral route and are responsible for high mortality rates in children.

These data, when evaluated together, can transcribe the characteristics and functioning of water bodies, an analysis that has become fundamental for sustainable development in a hydrographic basin. Currently, various programs for understanding hydrological behavior in river basins have emerged as a technique that broadens the reading of the factors occurring in these hydrological spaces (TUCCI, 2005).

REFERENCES

AL-RABEB, A. H; GUNAY, N. CEKIRGE, H. M. A hydrodynamic model for wind- driven and tidal circulation in the Arabian Gulf. **Appt. Math. Modelling**, Vol. 14, August.1990.

AMORIN, R. F.; SILVA, F. M.; Modeling the process of vulnerability to soil erosion using SPRING. Anais XIV Simpòsio Brasileiro de Sensoriamento Remoto, Natal, Brazil, April 25-30, 2009, INPE, p. 5073-5080.

ANA - National Water Agency. Water Resources Plan for the Piranhas-Açu River Basin (Report). Brasilia: Ministry of the Environment, National Water Agency, 2014.

ANDRADE, M. C. Energy production and modernization in the São Francisco valley. **Revista de economia politica**, vol. 4. N 1, January-March. 1984.

ARAÙJO, H. M. de et al. (Org.). Hydrology and hydrogeology: quality and availability of water for human supply in the coastal basin of the Sergipe River. In: VILAR, W. C.; ARAÙJO, H. M. (Org.). **Territory, environment and tourism on the coast.** Sao Cristovao: UFS, 2010. p. 168-188

ASSAD, L. P. de F.; MANO, M. F.; DECO, H. T.; TORRES JUNIOR, A. R.; LANDAU, L. Basics of computational hydrodynamic modeling and pollutant dispersion. Coppe, Federal University of Rio de Janeiro, Rio de Janeiro, 2009

BARBOZA, C.D.N.; PAES, E.T.; JANDRE, K.A.; MARQUES, JR., A.N. Concentrations and fluxes of nutrients and suspended organic matter in a tropical estuarine system: The Tinharé-Boipeba Islands Archipelago (Baixo Sul Baiano, Brazil). **Journal of Coastal Research**: Volume 30, p. 1197 - 1209. 2014

BAIRD, M. E; ADAMS, M. P; BABCOCK, R. C; OUBELKHEIR, K; MONGIN, M; WILD-ALLEN, K. A; SKERRATT, J; ROBSON, B. J; PETROU, K; RALPH, P. J; O'BRIEN, R, K; CARTER, A. B;

JARVIS, J. C; RASHEED, M. A. A biophysical representation of seagrass growth for application in a complex shallow-water biogeochemical model. **Ecological Modelling** 325 (2016) 13-27.

BASSANEZI, R.C. **Ensino-aprendizagem com modelagem matematica: uma nova estratégia**. Sao Paulo, SP: Editora Contexto, 2002.

BATISTA, S. S. EVALUATION OF THE EFFECTS OF ANTHROPIC ACTIONS ON THE COASTAL SYSTEM OF UBATUBA (SP) THROUGH ENVIRONMENTAL MODELING. **Dissertation**. University of Sao Paulo Institute of Energy and Environment Postgraduate Program in Environmental Science. 2016. 121p.

BEZERRA, I. S.; MENDONÇA, L. A. R.; FRISCHKORN, H. Self-depuration of watercourses: a Streeter-Pheips modeling program with automatic caiibration and anaerobiosis correction. Revista Escoia de Minas, n.2, v.61, p.249- 255, 2008.

BOTELHO, R. G. M.; DA SILVA, A. S. Bacia hidrografica e quaiidade ambientai. In: VITTE, A. C.; GUERRA, A. J. T. Refiexoes sobre a geografia fisica no BrasiLRio de Janeiro: Bertrand Brasii, 2004.

BRANCO, S. M. **Àgua: origem, uso e preservaçâo**. Moderna. Sao Pauio, 1993.

BRUNDTLAND. UN Documents Gathering a body of giobai agreements: Our Common Future, Chairman's Foreword. 1987. Available at:< http://www.un- documents.net/ocf-cf.htm>. Accessed on: June 16, 2016.

CARDOSO, C. A. DIAS, H. C. T.; SOARES, C. P. B.; MARTINS, S. V..Morphometric characterization of the hydrographic basin of the Debossan River, Nova Friburgo, RJ. **R. Àrvore**, Viçosa-MG, v.30, n.2, p.241-248, 2006.

CASTRO. C. F.A; SCARIOT, A. Water scarcity creates new injustice: water excision. **UNDP**. 2008. Available at:< http://www.pnud.org.br/gerapdf. php2id01=1067 Bariow M, Ciark> Accessed January 10, 2016.

CUNHA C. L. N.; MONTEIRO, T.; ROSMAN P.C. C. Two-dimensional modeling of non-conservative scalar transport in shallow water bodies. **Revista Brasileira de Recursos Hidricos** 2002; 7:120-9.

CUNHA, C. L. N.; ROSMAN, P. C. C.; FERREIRA, A. P.; TEÓFILO, C. N. M. Hydrodynamics and water quality models applied to Sepetiba Bay. **Continental Shelf Research,** v. 26, n. 16, p. 1940-1953, 2006.

CHRISTOFOLETTI, A. **Modeling Environmental Systems**. Sao Pauio: Edgard Biücher, 236 p. 1999.

DERISIO, J. C. **Introduction to environmental pollution control**. São Paulo: Editora Signus. 2007. 224p.

DUTTAA, D; WILSON, K; WELSH, W. D; NICHOLLS, D; KIM, S; CETIN, L. O. A new river system modelling tool for sustainable operational management of water resources. **Journal of Environmental Management** 121 (2013) 13 and 28.

FELGUEIRAS, C.A. Environmental modeling with treatment of uncertainties in geographic information systems: The geostatistical paradigm by indication. 2001. 215p. Thesis (Doctorate) - National Institute for Space Research, Sao José dos Campos.

FERNANDES, R. D. Formation of the Sand Banks at the mouth of the Amazon River. Thesis Ocean Engineering Program - Coastal & Oceanographic Engineering Area - COPPE/UFRJ. 2010

FERREIRA, R. A.; SILVA-MANN, R.; ARAGAO, A. G.; REZENDE, A. M. da S.; SANTOS, T. I. S.; SANTOS, P. L.; CARVALHO, S. V. A. Riparian areas in the Lower São Francisco region: occupation process and their recovery. In: LUCAS, A. A. T.; AGUIAR NETTO, A. O. (Org.). **Aguas do Sao Francisco.** Sao Cristovao: Editora UFS, 2011, v., p. 85-126.

FOLADORI, G. Sustentabilidad ambiental y contradicciones sociales. **Revista Ambiente & Sociedade** - Ano II - n 5 - 2o Semestre de 1999 19-34p.

GANJU, N. K.; KIRWAN, M. L.; DICKHUDT, P. J.; GUNTENSPERGEN, G. R.; CAHOON, D. R.; KROEGER, K. D. Sediment transport-based metrics of wetland stability. Geophysical Research Letters. Vol. 42. 2015. p 7992-8000.

GALLO, E; SETTI, A. F. F. Ecosystemic and communicative approaches in the implementation of territorialized Agendas for sustainable development and health promotion. **Revista Ciència e Saùde Coletiva**, 17(6):1433-1446, 2012.

GONÇALVES, C. U.; OLIVEIRA, C. F. de. Rio Sao Francisco: The waters flow to the market. **Boletim goiano de geografia**, Goiània, v. 29, n. 2, p. 113 - 125. 2009.

HAIDER, H.; ALI, W.; HAYDAR, S. Evaluation of various relationships of reaeration rate coefficient for modeling dissolved oxygen in a river with extreme flow variations in Pakistan. **Hydrological Process**. 2012.15 p.

JI, Z. G.HU, G.SHEN, J.WAN, Y. Three-dimensional modeling of hydrodynamic processes in the St. Lucie Estuary. **Estuarine, Coastal and Shelf Science** 73 (2007)188 and 200.

KANNEL, P. R.; KANNEL, S. R.; LEE, S.; LEE, Y.; GAN, T. Y. A review of public domain water quality models for simulating dissolved oxygen in rivers and streams. **Environmental Modeling & Assessment**, n.2, v.16, p.183-204, 2011.

KOSMINSKY, L. ZUFFO, A.C. O Nordeste Seco e a Transposigao do Rio Sao Francisco. APR./MAY./JUN. **Rev Integraçâo** Ano XV N^0 57.167-175. 2009. Available at: ftp://ftp.usjt.br/pub/revint/167_57.pdf> Accessed on September 28, 2015.

LACROIX, G; RUDDICK, K; GYPENS, N. LANCELO, C. Modelling the relative impact of rivers

(Scheldt/Rhine/Seine) and Western Channel waters on the nutrient and diatoms/Phaeocystis distributions in Belgian waters (Southern North Sea).**Continental Shelf Research** 27 (2007) 1422-1446.

LAGO, A. A. C. Stockholm, Rio, Johannesburg: Brazil and the three United Nations environmental conferences. Brasilia: Rio Branco Institute (IRBr)/ Alexandre de Gusmao Foundation (FUNAG) - Ministry of Foreign Affairs, 2006.

LARENTIS, D. G. Mathematical modeling of water quality in large basins: Taquari-Antas System-RS. 159f. Dissertation (Master's Degree in Water Resources and Environmental Sanitation) - Federal University of Rio Grande do Sul, Porto Alegre, 2004.

LEENDERTSE, J. J. Aspects of a computational model for long water wave propagation, Memorandum RH-5299-RR, Rand Corporation, Santa Monica. 1967.

LIMA, E. B. N. R. Integrated Modeling for Water Quality Management in the Cuiaba River Basin. 206 p. Thesis (Doctorate in Civil Engineering): Federal University of Rio de Janeiro, Rio de Janeiro. 2001.

LIU, Y.; GUPTA, H.; SPRINGER, E.; WAGENER, T. Linking science with environmental decision making: Experiences from an integrated modeling approach to supporting sustainable water resources management. **Environmental Modelling and Software**, v. 23, p. 846-858, 2008.

LILLEBO, A. I.; NETO, J.M.; MARTINS, I.; VERDELHOS, T.; LESTON, S.; CARDOSO, P.O.; FERREIRA, S.M.; MARQUES, J.C.; PARDAL, M. A. Management of a shallow temperate estuary to control eutrophication: The effect of hydrodynamics on the system's nutrient loading. **Estuarine, Coastal and Shelf Science** 65, 697-707. 2005.

MADEIRA, W. V. Sustainable Amazon Plan and uneven development. **Revista Ambiente e Sociedade**. Sao Paulo v. XVII, n. 3. p. 19-34. jul.-set. 2014.

MARTINS, D. M. F; CHAOAS, R. M; MELO NETO, J. O; MELLO Jr. Impacts of the construction of the Sobradinho hydroelectric plant on the flow regime in the Lower São Francisco River. **Journals. Bras. Eng. Agricola Ambiental**, v.15, n.9, p.1054-1061, 2011.

MEDEIROS, P. R. P; KNOPPERS, B; SOUZA, W. F. L; OLIVEIRA, E. N. Transport of suspended material in the lower Sao Francisco River (SE/AL), under different hydrological conditions. **Revista Braz. J. Aquat. Sci**. Technol, 2011, 15(1): 4253.

NASCIMENTO. W. M.; VILAÇA, M. O.. Hydrographic Basins: Planning and Management. **Electronic journal of the Association of Brazilian Geographers**. Três Lagoas, n. 7, p. 102-121, 2008.

NASCIMENTO, E. P. Trajectory of sustainability: from environmental to social, from social to economic. **Revista estudos avançados** 26 (74), 2012. 51-64p

OLIVEIRA, F. A; PEREIRA, T. S. R; SOARES, A. K; FORMIOA, K. T. M. Use of a hydrodynamic model to determine flow from level measurements. **RBRH**, PortoAlegre, v. 21, n. 4,p. 707-718, 2016.

PACHECO. M. I. N.; LIRA, F. J. A piscicultura no Baixo Sao Francisco: possibilidades e limites. Political economy of development. **Revista economia politica do desenvolvimento**. Maceió, vol. 1, n. 5, p. 67-95, May/Aug. 2009

PAIVA, R. C. D. COLLISCHONN, W. BRAVO, J. M. 1D hydrodynamic model for channel networks based on MacCormack's numerical scheme. In: **RBRH : Revista brasileira de recursos hidricos**. Porto Alegre, RS Vol. 16, n. 3 (jul./set. 2011), p. 151-161.

PAULA, S. M.; RAMIRES, I.; SILVA, F. O. TEODOSIO, T. K. C.; CAMPOS, K. B. O.; BRABES, K. C. S.; NEORÃO, F. J. QUALITY OF WATER FROM THE RIO DOURADOS, MS - PHYSICAL-QUIMICAL PARAMETERS AND

MICROBIOLOGY. **Evidência**, joaçaba v. 13 n. 2, p. 83-100, jui./dez. 2013 83

PEREIRA, S. B.; PRISKI, F. F.; SILVA, D. D.; RAMOS, M. M. Estudo do comportamento hidrológico do Rio Sao Francisco e seus principais afiuentes.

R. Bras. Eng. Agric. Ambiental, v.11, n.6, p.615-622, 2007

PEREIRA, S.P. Modeling the bacteriological quality of the coastal waters of Fortaleza (Northeast Brazil). Thesis (Doctorate) - Federal University of Ceara, 2012. 174 p.

PIERSON, D. **O Homem no Vale do Sao Francisco**. Rio de Janeiro: SUVALE. 1972.

PORTO, M. F. A. PORTO, R. L. L. Watershed management. **Estud. av.** 2008, vol.22, n.63, pp.43-60. Available at: http://dx.doi.org/10.1590/S0103- 40142008000200004 accessed in December 2015.

RAMINA, R. H. Concepçao uma estratégia robusta para Gestao dos usos Mùltiplos das Aguas Bacia Hidrografica Rio Sao Francisco. **CBHSF.** COMITÉ DE BACIA HIDROGRAFICA DO RIO SÃO FRANCISCO AGB - PEIXE VIVO. n. 14, 2014.

RATTNER, H. Sustentabilidade - uma viso humanista. **Revista Ambiente & Sociedade** - Ano II - no 5,2o Semestre de 1999. 233-240 p.

REBOUÇAS. A. C. Freshwater in the world and in Brazil. In: REBOUÇAS, A. C.; BRAGA, B.; TUNDISI, J. G. Aguas doces no Brasil - capital ecològico, uso e conservaçao. 2 ed. Sao Paulo: Escrituras Editora, 2002.

RIBEIRO, A. N; J CIRILO, J. A; DANTAS, C. E. O; SILVA, E. R. S. Characterization of flood formation in the Una River basin in Pernambuco: hydrological-hydrodynamic simulation. Revista Brasileira de Recursos Hidricos. vol. 20 no .2.2014.

RICCIARDONE, P. PEREIRA, O. DOS S.; PEREIRA, C. de S. S. Evaluation of the Self-depuration

Capacity of the Rio das Mortes in the Municipality of Vassouras/RJ. **Revista Eletronica TECCEN**, Vassouras, v. 4, n. 3, p. 63-76, set./dez., 2011.

RIGO, D. Flow analysis in estuarine regions with mangroves - measurements and modeling in Vitória Bay, ES. 2004. PhD Thesis in Ocean Engineering Sciences - Graduate Program in Ocean Engineering, COPPE/UFRJ, Rio de Janeiro, Brazil, 2004.

ROLLNIC, M.- Wave Dynamics and Circulation on the Coast of the Metropolitan Region of Recife - Implications for Sediment Transport and Coastal Stability. Thesis Defended by PPG-Oceanography/UFPE 2008.

ROSMAN, P.C.C. A computational system for environmental hydrodynamics. In: Silva, R.C.V. (Ed.), Numerical Methods in Water Resources - Volume 5. ABRH, 2001. p. 1-161.

ROSMAN, P. C. C. - Technical reference for sisbahia - base system for environmental hydrodynamics. COPPE/UFRJ, 2011. Available: www.sisbahia.coppe.ufrj.br Accessed: April 2016.

ROSMAN, P.C.C. SisBaHiA Technical Reference. COPPE/UFRJ. Rio de Janeiro-RJ. 2013.

ROSMAN, P. C. C., (Ed.). SisBaHiA® technical reference. COPPE - Federal University of Rio de Janeiro, March 2014, Available at: <http://www.sisbahia.coppe.ufrj.br/SisBAHIA_RefTec_V95. pdf> Accessed on: Jun. 2016.

SAMPAIO, A. F. P. Evaluation of the correlation between water quality and socioeconomic parameters in the Santos-Sao Vicente estuarine complex, through environmental numerical modeling. **Dissertation**. University of Sao Paulo Postgraduate program in environmental science - PROCAM. 2010.155p

SANTANA, D.P. Manejo Integrado de Bacias Hidrograficas. Sete Lagoas: Embrapa Milho e Sorgo, 2003. 63p.

SANTOS, R. G. Socio-environmental impacts on the banks of the São Francisco River: Result of the lack of consideration of the real area of influence. **GEOUSP** - Espago e Tempo, Sao Paulo, Special Issue, p. 81 - 91, 2009.

SANTOS, E. S. Hydrodynamic modeling and water quality in a pororoca region at the mouth of the Araguari-AP River. **Dissertation**. Fundagao Universidade Federal do Amapa conservacao internacional empresa brasileira de pesquisa agropecuaria instituto de pesquisas cientificas e tecnológicas do estado do Amapa. Pro-rectory for research and postgraduate studies Postgraduate program in tropical biodiversity. 2012. 114p.

SANTOS, J. W. B. Surface water salinity dynamics in the lower course of a coastal hydrographic basin. Thesis (Doctorate in Ecology and Biodiversity Conservation) - State University of Santa Cruz. 2014

SETTI, A. F. F.; BOGUS, C. M. Community Participation in an Intervention Program in an

Environmental Protection Area. Saùde Soc. v.19, n.4, p.946-960. Sao Paulo, 2010

SILVA, C. L.; MENDES, J. T. G. **Reflections on sustainable development**. Agents and interactions from a multidisciplinary perspective. Petrópolis: Vozes, 2005.

SILVA, J. A. **Direito Ambiental Constitucional**. 9. ed. Sao Paulo: Malheiros, 2011.

SILVA, T. R. Hydrodynamic analysis of the Auvioestuanno stretch of the Paraguaçu River, as a result of the operation of the Pedra do Cavalo hydroelectric plant. Dissertation. Master's Program in Environment, Water and Sanitation, Federal University of Bahia - UFBA. 2014.104p.

SHI, J. Z. A short note on the dispersion, mixing, stratification and circulation within the plume of the partially-mixed Changjiang River estuary, China. **Journal of Hydro-environment Research** 5 (2011) 111e126.

SHI, J. Z.; LU, L. F. A short note on the dispersion, mixing, stratification and circulation within the plume of the partially-mixed Changjiang River estuary, China. **Journal of Hydro-environment Research**. Volume 5, Issue 2, June 2011, p 111-126.

SOKOLOVAA, E; PETTERSSON, T. J. R; BERGSTEDTA, O; HERMANSSON, M; Hydrodynamic modelling of the microbial water quality in a drinking water source as input for risk reduction management. **Journal of Hydrology** 497 (2013) 15-23.

SOUZA, C. F; COLLISCHONN, W. ; TUCCI, C. E. M. Hydrodynamic simulation via remote data acquisition. In: XVII Simpòsio Brasileiro de Recursos Hidricos, 2007, Sao Paulo. Proceedings of the XVII Brazilian Symposium on Water Resources, 2007.

SOUZA, J. F. A.; OLIVEIRA, L. R.; AZEVEDO, J. L.L.; SOARES, I. D.; MATA, M. M. A review of turbulence and its modeling. **Rev. Bras.**

Geof.[online]. 2011, vol.29, n.1, pp.21-41. Available at:<

http://www.scielo.br/pdf/rbg/v29n1/02.pdf> Accessed on: <January 13, 2016>.

SOUZA, M. M; GASTALDINI, M. C. C. Assessment of water quality in watersheds with different anthropogenic impacts. **Eng. Sanitaria Ambiental**, Rio de Janeiro ,v. 19, n. 3, p. 263-274, Sep. 2014 .

SPAREMBERGER, R. F. L; MARQUES, C. A. M. Hollow men and the environment: Sustainable Development for Whom? **Revista direito em debate**. Year XXIV n⁰ 43, Jan.-Jun. 2015 - p. 3-26

TUCCI, C. E. M.; HESPANHOL, I.; CORDEIRO F. O. **Water management in Brazil: an initial assessment of the current situation and prospects for 2025**. Brasilia, GWP, 2000. 165 p.

TUCCI, C. E. M. Development of water resources in Brazil. **REGA**, v.2, n.2, dec. 2005

TUCCI, C. E. M. **Hidrologia: ciência e aplicação**. 4.ed, Porto Alegre: Editora UFRGS/ABRH, 2007, p. 943.

TORTORA, G. J.; FUNKE, B. R.; CASE, C. L. **Microbiology.** 8. ed. Porto Alegre: Artmed, 2005. 894 p.

UNEP. United Nations Environment Program. Emerging Issues of Environmental Concern. UNEP Frontiers 2016. Available at:< https://uneplive.unep.org/media/docs/assessments/UNEP_Frontiers_2016_repo rt_emerging_issues_of_environmental_concern.pdf. Accessed on: June 19, 2016.

UNESP. Declaration of the United Nations Conference on the Human Environment. Available at:<http://www.unep.org/Documents.Multilingual/Default.asp?DocumentID=97 &ArticleID=1503&l=en>. Accessed on: June 15, 2016.

USACE. HEC-RAS River Analysis System: Hydraulic Reference Manual, Version 4.1. U.S. Army Corps of Engineers, Hydrologic Engineering Center. 2010.

USACE-HEC. Hydrologic Modeling System, HEC-HMS v3.5 - User's Manual. US Army Corps of Engineers, Hydrologic Engineering Center. 318 p., 2010.

VEIGA, B. E.; MAGRINI, A. Water resources, climate change and adaptation: proposals for Brazil in the light of the European Union. In: Simpòsio brasileiro de recursos hidricos, agua - desenvolvimento económico e socioambiental, v. 20, 2013, Bento Gonçalves. Proceedings. Bento Gonçalves: ABRH, 2013.

XAVIER, A. G. Analise da hidrodinâmica da baia de Todos os Santos (BA). 205 f. Thesis (Doctorate in Science in Ocean Engineering) - Federal University of Rio de Janeiro, COPPE, Rio de Janeiro, 2002.

WWDR (World Water Assessment Program). United Nations World Water Development Report: Water for a sustainable world. UNESCO. 2015.   8p. Available at:<

http://www.unesco.org/new/fileadmin/MULTIMEDIA/HQ/SC/images/WWDR201 5ExecutiveSummary_POR_web.pdf>Accessed on: 31/06/2016.

ZELLHUBER, A; SIQUEIRA, R. Rio Sao Francisco em descaminho: Degradação e revitalizaçao. In: Rio Sao Francisco transposition environmental degradation alternatives. **Cadernos do CEAS.** Salvador Centro de Estudos e Açao Social, n^0 227, July-September, 7-32p. 2007.

ZHOU, Y; GUO, S; CHONG-YU, XU; LIU, D; CHEN, L; WANG, D. Integrated optimal allocation model for complex adaptive system of water resources management (II): Case study. **Journal of Hydrology** 531 (2015) 977-991.

CHAPTER 2: CHARACTERIZATION OF WATER QUALITY AT THE MOUTH OF THE SÂO FRANCISCO RIVER

SUMMARY

Water reflects the environmental conditions of a river, so knowing its quality broadens our understanding of the hydrodynamics of this ecosystem and makes it possible to detect anthropogenic actions that affect human activities and the maintenance of ecological processes. The aim of this study was to assess the quality of the water at the mouth of the River San Francisco and its anthropogenic effects. The mouth of the São Francisco River is located between the states of Sergipe and Alagoas. It is a fluvial and marine plain made up of varzeas and river and marine terraces. It represents an area of extreme natural beauty and economic importance for the local population and is part of the geographical division of the lower São Francisco. Using a boat located in the municipality of Brejo Grande (SE) and with the aid of a Garmin e-trex digital GPS, latitude and longitude information was recorded in UTM to map the water collection points. Using the acoustic doppler profiling method (ADCP), a bathymetry section was carried out at the mouth of the River San Francisco to determine flow and depth. To check water quality and degradation processes, the sampling points were identified through measurements with a HI9828 Multiparameter probe during the months of February, March, April, May, September, November 2015 and March 2017. The probe recorded salinity, electrical conductivity, turbidity, dissolved oxygen, total solids, temperature and pH. For drinking water, water was collected from households to check pH, sodium and chlorides. Interviews were conducted with residents of the Brejo Grande/SE and Saramèm regions to ascertain their perception of salinity. The quantitative and qualitative data was then processed using statistical software and compared with CONAMA Resolution 357/2005 and Ordinance MS 2914/2011. The results for pH ranged from 6.80 to 9.47, with the highest value occurring in February 2015. Dissolved oxygen values in the area evaluated ranged from 4.68 to 9.27 ppm. The total dissolved solids values ranged from 7.0 mg.L^{-1} to 7522 mg.L^{-1} . The highest concentration of total dissolved solids recorded occurred at Point 1 in November, a dry period with a sizig tidal regime and a 2.2m rise in the water level. The turbidity values recorded were between 2.9 NFU and 47.07 NFU. Electrical conductivity values ranged from 10.00 (µS/cm^{-1}) to 13130.00 (µS/cm^{-1}), with the highest value recorded at Point 7 and the lowest at Point 38. The average salinity concentration ranged from 0.17 to 28.87 (‰), with the highest values recorded at Points 7 and 11 located on the edge of the municipality of Brejo Grande/SE and Points 39, 40 and 43 located towards Piaçabuçu in the state of Alagoas. The pH in drinking water between the municipalities of Brejo Grande/SE and the village of Saramèm showed average values within the limit established by Ministry of Health Ordinance 2914/2011. In the evaluation of the sodium and chloride parameters, the average values obtained in the month of May 2015 were between

1962.85 mg.L^{-1} and 2938.62 mg.L^{-1} . In response to the semi-structured interviews, residents reported that the change in the taste of the water was perceived during high tide. It can be concluded that the salinity concentrations were significant and signaled that the tide period acts to increase salinity. Salinity is a worrying factor because it directly compromises water use and affects the biota in the region. It is suggested that further research be carried out in the region to monitor the saline wedge and its potential risks for the population of the lower São Francisco.

Key words: Water quality. Salinization. Anthropic actions.

2.2 Introduction

Water reflects the environmental conditions of a river, so knowing its quality broadens our understanding of the hydrodynamics of this ecosystem and makes it possible to detect anthropogenic actions that affect human activities and the maintenance of ecological processes.

According to Oliveira, Campos and Medeiros (2010), verifying that a given body of water has satisfactory conditions is to ensure its uses, but it is necessary to carry out physical and chemical characterization, that is, to assess its quality and compare its data according to the classification of CONAMA Resolution 357l05 and Ordinance MS 291412011. The evaluation process depends fundamentally on the choice of representative parameters to identify the degradation profile (OLIVEIRA, CAMPOS and MEDEIROS, 2010).

Due to population and economic growth, the need for energy generation, food and land use has become intense, causing changes to natural resources. These changes reflect impacts of great magnitude that can directly or indirectly affect water quality and limit its use.

The mouth of the River San Francisco is an area of extreme natural beauty and economic importance for the local population. It is part of the geographical division of the lower San Francisco, between the municipalities of Brejo Grande and Piaçabuçu. In the lower São Francisco region, the last dam corresponding to the Xingó Hydroelectric Power Plant is located on its banks, which has the greatest potential for generating energy in the Northeast.

The process of damming a river course interferes with the pulses of natural flows, altering the dynamics of its waters and reducing its flow along its course. The reduction in the flow of the San Francisco River has already been signaled as unsustainable for guaranteeing its uses. Martins et al, (2011) add that the flow regime determined by the regularization of the river for the purposes of energy generation constitutes points of conflict between public supply, irrigation and environmental flow.

Santos et al (2013) analyze that, due to the regularization of the flow, the saline wedge from the coastal waters already proliferates around 10 km inland and the influence of the tide extends beyond this limit. The presence of the salt wedge in the course of a river causes salinization of its waters, adding salts that limit its use for public supply, irrigation and ecological processes.

The salt wedge is carried into the river by the oscillations of the tides. As the sea water is denser than the river water, the movement in and out promotes this displacement. According to Peter and Huber (2008) the salt wedge moves back and forth with the daily rhythm of the tides, so it enters at high tide and recedes at low tide.

With the regularization of flows, the river's action on the sea is slower, allowing the salt wedge to move upstream. Cavalcante et al (2017) add that the discharge of the river and the tidal currents influence the circulation and transport of salt in the estuary of the River San Francisco, especially at high tides.

In view of the above, the aim of this study was to characterize the quality of the water in the area of the mouth of the River San Francisco, its anthropogenic actions and the possible limitation of its uses.

2.   2 Methodology

2.1.1 Characterization of the study area

The hydrographic basin of the São Francisco River comprises approximately 640,000km^2 of drainage area, whose main river is 2,697km long with an average flow of 2,843.6m^3/s (CBHSF, 2016). The source of the São Francisco River is located in the Serra da Canastra in Minas Gerais, running through the states of Bahia, Pernambuco, Alagoas, Sergipe, Goias and the Federal District, linking Brazil from the Southeast to the Northeast, representing 7.5% of the country (AGUIAR NETTO et al 2011).

According to the National Water Agency (2005), this basin is divided into four physiographic regions: Alto, Mèdio, Submédio and Baixo Sao Francisco (Figure 1), for planning purposes these areas were subdivided into thirty-four small basins, and 12,821 micro basins in order to delineate the main rivers in the region by sections.

The general characteristic of the physiographic regions corresponds to 16% of the basin's area, starting between the source of the main river in the Serra da Canastra in Minas Gerais and the confluence with the Jequitai River. The main hydroelectric dams in this location are Três Marias, Retiro Baixo and Queimado (CBHSF, 2016). The middle São Francisco is the largest representation of the São Francisco River basin, occupying 63% of its territory, and the main dams are Sobradinho and Rio das Fèmeas. The sub-middle reaches 17% of the basin's area from the Sobradinho dam to the Xingò dam, the main cities are Petrolina (PE), Serra Talhada (PE), Juazeiro (BA) and Paulo Afonso (BA), including the Paulo Afonso I, II, III and IV, Xingò, Itaparica and Moxotò dams (CBHSF, 2016).

The smallest physiographic region of the basin is represented by the lower São Francisco, corresponding to 4% of the area of the hydrographic basin between Xingò and its mouth in the Atlantic Ocean. Located between the parallels of 8^0 and 11^0 South Latitude and the meridians of 360 and 390 West Longitude, with a territorial expansion of 23,546.32km^2 it represents the westernmost portion of the Sao Francisco hydrographic basin, covering the states of Bahia, Pernambuco, Sergipe and Alagoas, culminating in its mouth (CAVALCANTI, 2011; LIMA et al 2010). According to Cavalcanti (2011) this area stretches from the Xingó hydroelectric power station (Canindé do Sao Francisco/SE) to the mouth region, corresponding to an extension of approximately 210 km.

The mouth of the São Francisco River, located between the states of Sergipe and Alagoas, has a fluvial-marine plain consisting of varzeas and fluvial and marine terraces formed by Quaternary deposits and rounded hills carved out of sedimentary rocks and surrounded by coastal tablelands (SANTOS, 2010). The Foz region is subject to a meso-tidal regime, with semi-diurnal tides (two low tides and two high tides).

According to Valente et al (2011), the estuary has a total area of 100km^2 with coastal plain environments and direct influence from ocean waters. The flood plains have had their hydrological dynamics compromised by the construction of dams for electricity generation.

According to Fontes (2015), the semiarid region of the lower São Francisco contains mostly intermittent tributary rivers, while perennial tributaries only exist in the coastal region, with the municipality of Propria/SE being positioned between the semiarid and the humid regions (CARVALHO and FONTES, 2007).

Figura 1: Map of the Sâo Francisco River Basin, its physiographic regions and, in particular, the Lower Sâo Francisco.

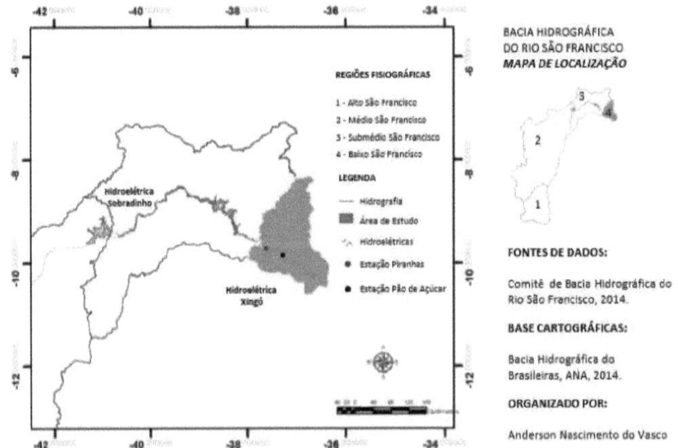

Figura 2: Lower Sâo Francisco region, in particular the municipalities of Propria/SE, Santana de Sâo Francisco, Neópolis/SE, Ilha das Flores/SE, Brejo Grande/SE, Penedo/AL and Piaçabuçu/AL.

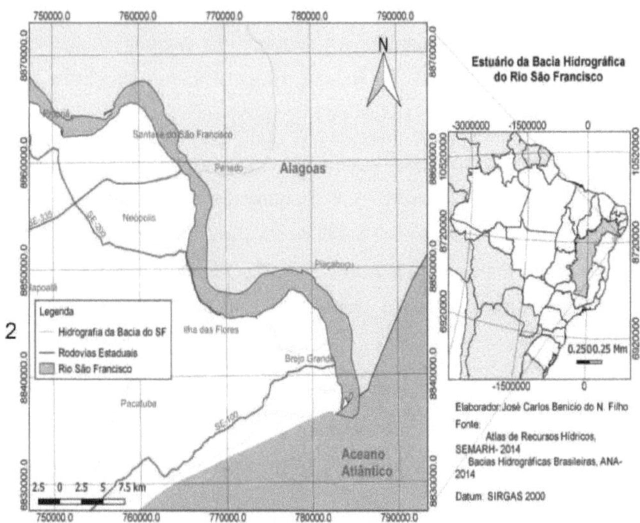

2.1.2 Abiotic characteristics of the study area

The lower course of the São Francisco River has a "canyon" in the remobilized massifs of the pediplano in the city of Paulo Afonso (BA), extending, in the form of waterfalls in a rocky bed with gravels, for about 100 km to the vicinity of the city of Pao de Açùcar (AL), recently interrupted by the lake of the Xingó Hydroelectric Power Plant dam (CAVALCANTI, 2011).

Geologically, it is part of the Borborema province. This province is characterized by the existence of metamorphic rocks originating from the clash between tectonic plates that occurred in the Brasiliano cycle, sediments were added in the Jurassic and Cretaceous periods to form the geological structure of the area (BRASIL, 2002).

More than 80% of the stretch between Paulo Afonso (BA) and Propria (SE) is made up of rocks from the Precambrian Indiviso, while the remaining 20%, from Propria to the mouth, is made up of Cretaceous rocks from the Coruripe Indiviso sub-group in the same way and in the same proportion (10%) as the Quaternary unconsolidated deposits. Finally, the remaining 10% is made up of Tertiary and Quaternary sediments from the Barreiras Formation (BRASIL, 2006).

The geomorphology is characterized by three morphostructural domains, the first of which is considered to be a sedimentary deposit represented by plateaus and trays made up of predominantly unconsolidated sediments extending along the marginal areas of the São Francisco river and the coasts of Sergipe and Alagoas. The features found reflect the deposition of sediments accumulated in marine, fluvial, marine fluvial, eolian and colluvial environments (BRASIL, 2002).

The second morphostructural domain is called the remnant of the fold roots, located in the central

29

portion of the Lower São Francisco River, approximately between the towns of Porto da Folha and Propria. This domain is characterized by an alignment of sub-sequent and perpendicular ridges and valleys, the result of differential dissection and the wearing away of folded structures and occasional exposures of their basement (CAVALCANTI, 2011).

The third domain of Remobilized Massifs occupies the largest area of the lower São Francisco, stretching from Porto da Folha (SE) to Paulo Afonso (BA). It is characterized by forms resulting from the erosion of arched blocks displaced by tectonic reactivations. Particularly noteworthy are the occurrence of inselbergs, ridges and deep furrows in the diaclastic zones. Flat topographies appear in regions protected from the retreat of fluvial erosion (CAVALCANTI, 2011; BRASIL, 2002).

The soil is characterized by planossolo, neossolo, argilossolo, espodossolo, latossolo, and the main mineral reserves exploited include ornamental granite and sylvinite (BRASIL, 2016). In the interior of the islands, we can see ancient evidence of sandy points (spits), immobilized by the spread of the coastline, which shelter halomorphic soils in the innermost portions under the influence of the tides, with characteristic mangrove vegetation (CARVALHO and FONTES, 2007).

In the region of the mouth of the São Francisco River, there is intense interaction between fluvial and marine processes. The main contribution of the river is the supply of fresh water (measured by the flow of water) and the load of sediment. The ocean participates with salt water, the movement of which promotes coastal dynamics through coastal currents, wave behavior and the tidal regime. When the São Francisco River was a natural river, not controlled by man through dams, its liquid and solid discharges imposed themselves on the mouth region, largely determining the morphological features and the distribution of erosion and sedimentation in the adjacent coastal zone (CBHSF, 2013).

The climate in the lower São Francisco gradually changes from arid to semi-arid in the interior of the continent and from sub-humid to humid in the coastal region where the mouth is located. Between Propria (SE) and the coastal zone the average temperature is between 23 and 26° C and in normal situations is marked by a well-defined rainy season with abundant rainfall (FONTES, 2015). In terms of rainfall, there are two distinct seasons: the rainy season between April and August and the dry season between September and March (MEDEIROS, 2003).

The Sao Francisco is one of the few perennial rivers in the working area and is used for various social and economic purposes, such as: supplying water to urban populations (including the city of Aracaju), diluting domestic effluents, supplying irrigated agriculture, planting short-cycle crops, aquaculture, ecotourism, navigation and exploiting hydroelectricity through the Xingó Plant, owned by the Companhia Hidrelétrica do Sao Francisco - CHESF (BRAZIL, 2002).

The water supply in this region is conditioned by regular changes in the flow, with continuous values interfering in the seasonality process due to the influence of the dams. Vasco (2015) shows that the construction of reservoirs has drastically changed the seasonality of the lower São Francisco, making it impossible to adopt flows that are close to the river's natural flow. For the author, the adoption of seasonal values is the best way to represent the natural conditions of a river. Vasco (2015), during his research, presented flow values of 1,100 m^3/s, currently the recorded flow is around 900 m^3/s, with indications of a reduction to 600 m^3/s.

It should be noted that the lower Sao Francisco is home to the Xingó hydroelectric plant, belonging to the Companhia Hidrelétrica do Sao Francisco (CHESF), with an installed power generation capacity of around 3,000 MW. Much of this energy is exported to the large urban centers of the Northeast (BRASIL, 2002).

2.1.3 Biotic characteristics of the study area

The vegetation cover of the lower Sao Francisco region includes fragments of various ecosystems. The caatinga covers the western part of the hydrographic basin, the Atlantic forest the eastern part, and as the São Francisco River reaches its mouth there are pioneer fluvial-marine formations (FERREIRA et al 2015).

The distribution of vegetation in the Lower San Francisco varies from the interior to the coastal zone. The extensive areas of caatinga represent the original layer in the Lower São Francisco region and constitute a xerophytic type of vegetation, i.e. with functional adaptations against the lack of water, developed as a result of the low level of rainfall in the semi-arid region (FONTES, 2015).

There are two physiognomic variations of this type of vegetation: the dense arboreal caatinga and the open arboreal caatinga. The former can be characterized structurally by having a dense arboreal layer with species that vary from 8 to 10 m in height; a layer formed by thorny evergreen shrubs and, finally, a seasonal herbaceous layer. The second physiognomic variation has a woody cover with an open structure and low size, leaving the grassland layer exposed. This type of caatinga can be found in natural conditions in areas with a markedly dry climate and lithic soils (BRASIL, 2002).

According to Ferreira et al (2005) the most common tree and shrub species in this region are *Poincianella pyramidalis* (Tul.) L.P. Queiroz (catingueira), *Aspidosperma pyrifolium* Mart. (pear tree), *Commiphora leptophloeos* (Mart.) J.B. Gillett (imburana-de-cambão), *Bauhinia Cheilantha* (Bong.) Steud. (mororò), *Jatropha mollissima* (Pohl) Baill. (pinhao-bravo), *Cereus jamacaru* DC. (mandacaru), *Pilosocereus gounellei* (F.A.C. Weber) Byles & G.D. Rowley (xique-xique), *Schinopsis brasiliensis* Engl. (braùna), *Libidibia ferrea* (Mart. ex Tul.) L.P. Queiroz (pau-ferro), *Ziziphus joazeiro* Mart. (juazeiro), *Spondias tuberosa* Arruda (umbuzeiro) and *Myracrodruon urundeuva* Allemao (aroeira).

In the area in question, the main types of vegetation are semideciduous seasonal forest, mangroves and coastal vegetation (BRASIL, 2006). In the Parapuca and Poço canals there is a greater development of mangroves, which occupy an area of 21.68 km^2. *Rhizophora mangle* is the dominant species, adapting well to this environment because it has anchor roots that allow it to attach itself to physically unconsolidated sediments (CARVALHO and FONTES, 2007). Restingas and mangroves are two types of environment that appear on the coasts of Sergipe and Alagoas.

Aquatic plants, with species typical of freshwater environments, such as *Montrichardia arborecens*, *Eichhornea sp.* and *Salvinia sp.*, present at some points, as well as mud banks occupied by individuals of the genera *Crenea* and *Juncus* complement the vegetation landscape of the river course (SEMENSATTO JUNIOR, 2006).

The Lower Sâo Francisco region also has a high diversity of endemism. According to Sigrist and Carvalho (2008), endemics are areas characterized by the presence of species with a restricted distribution. Endemic species are important for the conservation of environments. 79 endemic species have already been catalogued in the lower Sâo Francisco area, 59 (10.8%) from the Caatinga and 20 (3.6%) from the Atlantic Rainforest (FERREIRA et al 2015).

Migratory birds use the area around the mouth of the Sâo Francisco River to feed and rest when they are passing along the Brazilian coast. Migratory species include the little booby (*Puffinus puffinus, Procellaridae*); the fish eagle (*Pandion haliaetus, Pandionidae*); representatives of the Charadriidae family, the black-tailed godwit (*Pluvialis squatarola*) and the banded godwit (*Charadrius semipalmatus*); the representatives of the Scolopacidae family, white-backed curlew (*Limnodromus griseus*), grey-crowned curlew (*Numenius phaeopus*), spotted curlew (*Actitis macularius*), solitary curlew (*Tringa solitaria*), great yellow-legged curlew (*Tringa melanoleuca*), yellow-legged curlew (*Tringa flavipes*), turnstone curlew (*Arenaria interpres*), red-capped curlew (*Callidris canutus*), white curlew (VALENTE et al, 2011).

In the lower Sâo Francisco there are 33 freshwater fish species, 7 introduced, 14 marine and 1 hybrid species (Tambacu) (COSTA, 2003). In addition to fish, it is also possible to see reptiles and mammals, but some species are no longer so common to see. The jaguar, for example, is extinct in the Foz do Velho Chico region.

Biodiversity in the coastal region is recorded by the presence of APAs (Environmental Protection Areas), which are defined as extensive areas, with a certain degree of human occupation, endowed with natural, aesthetic and cultural attributes that are important for the quality of life and well-being of the population (MMA, 2011). The Piaçabuçu AAPA covers 18,000 hectares, with flooded areas, restinga forests and a mosaic of plant and animal species. It consists of a region characterized by the presence of dunes and sprawling grasses. The dunes are a 19 km long sandy cordon that runs along the beach from Pontal do Peba to the mouth of the Sâo Francisco river, varying in width from 700m to 800m, bordering the opposite side of the continent (CABRAL, AZEVEDO JUNIOR and LARRAZABAL, 2006).

2.1.4 Socio-economic characteristics of the study area

The population distribution of the Lower San Francisco corresponds to approximately 2,421,150 inhabitants, of which 62.54% (1,514,119 inhabitants) belong to the states of Alagoas and Sergipe. According to the 2010 Census, the population of these two states residing on the banks of the Lower San Francisco corresponds to a total of 442,728 inhabitants (BRASIL, 2013).

The household situation in the São Francisco River basin has shown a growing trend in the proportion of inhabitants in the urban sector, with an increase from 69% in 1991 to 75.9% in 2000 (ATLAS BRASIL, 2011). However, the reality of the riverside municipalities analyzed shows that the urbanization rate was 52.87%, which is lower than the average growth rates achieved over the last twenty years (BRASIL 2013). Factors such as education, sanitation and health are essential for the growth of the population and the lack of these services hinders the region's economic growth.

According to (BRASIL, 2006), the lower São Francisco represents 10.7% of the population of the São Francisco river basin. Most of the municipalities do not have a population of more than 100,000 inhabitants, and the population growth and socio-economic characteristics can be seen in Table 1. The main socio-economic activities include agriculture, livestock farming, fishing and aquaculture.

According to the 2010 Census, the population of Sergipe and Alagoas living on the banks of the Lower Sâo Francisco corresponds to a total of 442,728 inhabitants. In terms of basic sanitation, the Baixo Sâo Francisco region reproduces what can be seen in many areas of the country, i.e. unsatisfactory sanitation.

It is understood that basic sanitation measures can improve the living conditions of the rural population, as well as family farming communities. Among these measures are sanitary sewage services, which can be defined as the set of works and installations designed to collect, transport, remove, treat and finally dispose of the community's wastewater in an appropriate manner from a sanitary point of view (IBGE, 2011). Inadequate or absent basic sanitation is an issue related to the poverty of a community and the risk of various diseases.

The biggest deficits in the state, both in terms of coverage and quality of the drinking water supply and sewage services offered, are to be found in small communities in isolated rural areas. Sanitation in the municipalities of the lower Sâo Francisco is carried out by Companhia de Saneamento de Alagoas - CASAL/AL and Companhia de Saneamento de Sergipe - DESO (SE). According to Figures 3 and 4, access to drinking water in the municipalities is directly proportional to the number of inhabitants in the region.

Araùjo (2015) argues that, in the lower Sâo Francisco region, there has been an increase in the percentage of the population living in households with piped water, but there are still municipalities with little access to it.

Figura 3: Lower Sâo Francisco, in particular the municipalities of the state of Sercjipe and the relationship with access to water.

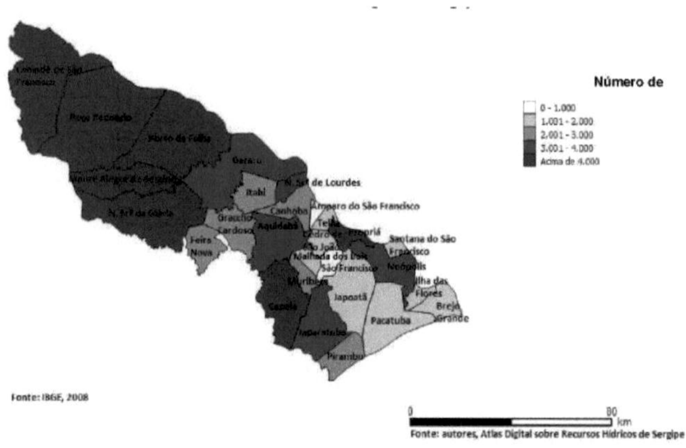

Source: Map drawn up by José Emilio de Jesus

Figura 4: Baixo Sâo Francisco, highlighting the municipalities in the state of Alagoas and their relationship with access to water.

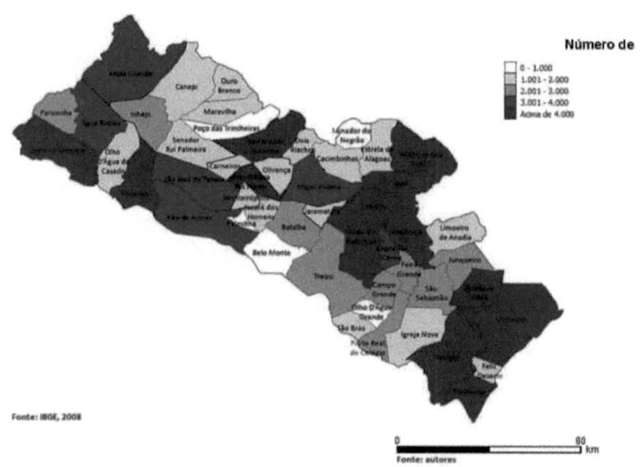

Source: Map drawn up by José Emilio de Jesus (2017)

In the Lower San Francisco region, the level of education is still low and the schools, besides being few in number, are mostly located in urban areas. The health situation in this area is quite precarious, with an insufficient number of hospitals and emergency rooms. However, the region has several tourist attractions such as canyons, museums, historical centers and the rich biodiversity distributed all the way to Foz (Figures 5 and 6).

In terms of population, the demographic density of the lower Sâo Francisco region is fluctuating, i.e. in the 2010 period, the population growth rate in the area was slightly higher than in 2000, with the population growing by around 10.4% from 1991 to 2000 and 8.3% from 2000 to 2010 (ARAÙJO, 2015). According to Figure 5, the most populous municipalities that belong to the lower Sao Francisco and are part of the state of Sergipe and directly linked to the course of the Sao Francisco River are: Canindé de Sao Francisco, Poço redondo, Porto da Folha and Propria and the least populous Amparo de Sao Francisco and Telha.

Figure 5: Lower Sao Francisco, highlighting the municipalities in the state of Sergipe and the number of inhabitants.

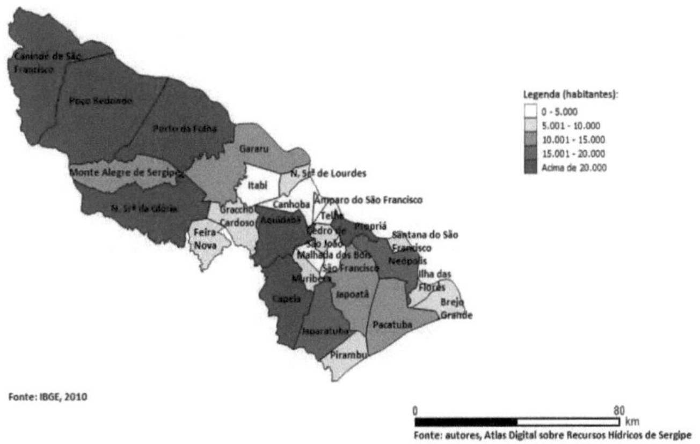

Source: Map drawn up by José Emilio de Jesus (2017)

In the region of the state of Alagoas, the most populous municipalities connected to the lower São Francisco and forming part of the course of the São Francisco River are: Delmiro Gouveia, Piranhas, Pao de Açúcar, Traipu, Igreja Nova and Penedo, while the less populous Olho D'agua Grande and Feliz Deserto are not bathed by the São Francisco River (Figure 6).

A summary of the main socio-economic characteristics developed and future prospects is shown in Table 1. Among the information, the expansion of the irrigated area for the 2020 period shows a strong growth trend, for Jesus and Gomes (2012) the lower Sao Francisco is characterized by being a region of large state and private investments directed, respectively, to the hydroelectric sector and to the modernization of agriculture with irrigated perimeters.

Fig. 6: Lower Sao Francisco, highlighting the municipalities in the state of Alagoas and the number of inhabitants.

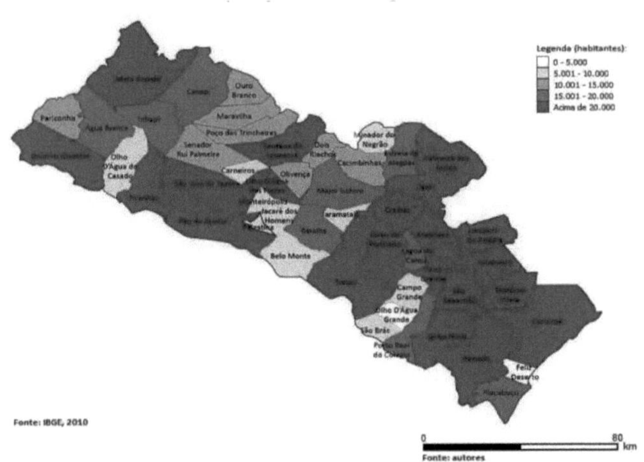

Source: Map drawn up by José Emilio de Jesus (2017)

Table 1: Main socio-economic characteristics of the lower Sâo Francisco region.

Features	2004	2006	2016
Population (inhab)	1.372.735	1.624.388	2.095.123
Urbanization (%)	51	51	-
Water Supply (%)	82,4	82,4	93 (Target by 2023)
Sewage collection (%)	23,4	23,4	76 (Target by 2023)
Garbage collection (%)	87,7	87,7	95 (Target by 2023
Irrigated area (ha)	34.681	34.681	70,000 (Target up to 2020)
HDI	0,364 a 0,534	0,364 a 0,534	0,538

Source: PRHBHSF[3] (2004, 2006 and 2016).

1.2.5 Environmental impacts

The environmental impacts affecting the lower Sâo Francisco come from anthropogenic actions, resulting from human use and occupation of the land. Population growth along this area also leads to environmental losses and the lack of existing information, especially hydrological, prevents many of its impacts from being properly assessed.

However, it is possible to detect some impacts present in the Lower Sâo Francisco environment: erosion and sedimentation caused by the occupation of the varzeas with the construction of dykes, another process was the construction of dams upstream. These block the passage of solids and,

---

3 Water resources plan for the Sâo Francisco river basin.

on the other hand, alter the hydrological cycle by regulating flows (IPEA, 1992).

Due to the regularization of flows, the natural fertilization of the varzeas is interrupted, necessitating the use of synthetic fertilizers and pesticides in these places. Analyses of water quality in tributaries have shown total phosphorus levels above the acceptable standard. Brito et al (2015) recently detected, in research on the Betume River, one of the tributaries of the Sâo Francisco River and located in the lower Sâo Francisco region, three active ingredients used in agricultural practices in the Betume irrigated perimeter region: chlorpyrifos, tebuconazole and tetrabuconazole, with their respective trade names Colosso, Nativo and Domark. It is noteworthy that the values were not above the permitted standards, but were indicative of the percolation of these chemical elements.

Currently, one of the widely publicized impacts is the salinization of the waters caused by the displacement of the salt wedge. According to Medeiros (2014), salinity is an important ecological factor in estuarine environments, due to the osmotic stress it causes organisms. Under natural conditions, the type and extent of saline intrusion in an estuarine environment depends on oceanographic forces (tides, waves, winds). The construction of cascade dams along the main course of the river has eliminated inter-annual variability. As a result of this loss of annual variability in freshwater flow, saline intrusion, in the form of a salt wedge, enters the mouth of the São Francisco River (MEDEIROS et al 2014).

Deforestation for agriculture and aquaculture is also considered a factor in environmental degradation. The mangroves are being exposed to the direct action of waves and the mudflats are being vigorously silted up, which has caused the death of well-developed woodland plants. Deforestation has been observed in the Parapuca channel in its closest part to the San Francisco, in the topographically higher areas, for the installation of coconut plantations. In addition to the Parapuca, small areas of deforestation for the construction of fish ponds can be seen near the village of Ponta dos Mangues (SEMENSATTO JUNIOR, 2006).

The lack of sewage treatment and the correct disposal of solid waste cause a loss of water quality in the lower São Francisco. With regard to sanitary sewage treatment, the presence of intermittent rivers makes it difficult to dilute effluents, and with regard to water supply, the absence of water sources with guaranteed quality and quantity makes it difficult to serve the population (BRASIL, 2006).

With the reduction in water quality, access to water resources is reduced, enabling conflicts over the use of water to the quality standards that meet its uses (BRASIL, 2006). Figure 7 shows that the irrigation sector is the largest allocator of water in the Lower San Francisco region.

Figure 7: Consumption flows for the different consumptive uses in the lower São Francisco region.

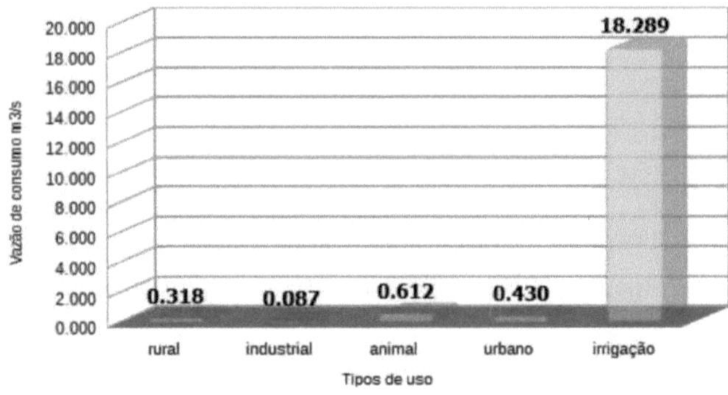

Source: ANA (2013 and 2016)

Finally, it is believed that one of the biggest conflicts on the San Francisco River today, the source of significant environmental impacts, is the lack of synchrony between the operating regime of the hydroelectric plants and the natural conditions characteristic of the river, which has manifested itself, for example, the disappearance of annual seasonality (which is responsible for the rhythms of endogenous ecosystems) and weekly, daily and hourly fluctuations (which are characterized by unpredictable river conditions), with negative impacts on navigation and all traditional social activities associated with river rhythms, such as artisanal fishing and agriculture in the varzeas (BRASIL, 2016).

There are indications that this region is undoubtedly the most hydro-environmentally vulnerable in the entire stretch of the São Francisco river basin (ANA, 2003). According to Lima et al (2010), this region has suffered impacts over the years which have compromised the use of natural resources and the performance of activities such as navigation, irrigation, tourism, energy generation, aquaculture, urban and industrial supply and fishing.

Martins et al (2011) also point out that the dams have caused direct effects on the lower São Francisco, including a reduction in the number of fish and invertebrate species, a reduction in the levels of sediment and nutrient deposition on the floodplain, impediments to navigation, changes in the biophysical processes of the estuaries, a reduction in underground recharge and compromised water availability for multiple uses.

1.2.6 . Different perspectives on the evaluated region

In a more targeted analysis of land use and occupation in the territories evaluated, they proposed scoring fragile areas, land use and qualitative information collection points. The area evaluated is located in the region of the mouth of the River São Francisco between the municipalities of Brejo Grande (SE) and Piaçabuçu (AL). The municipality of Brejo Grande (SE) is located on the banks of the São Francisco and is the last town in the state of Sergipe in the river basin. It is part of the rich and complex ecosystem of the mouth of the São Francisco.

The profile of land use and occupation in the lower Sâo Francisco is described in Figure 8, in the region of Brejo Grande/SE seasonal forest, crops on exposed soils, mangroves, dunes and clogged areas. Much of the land along the banks of the river Sâo Francisco is bare of vegetation between Brejo Grande/SE, Ilha das Flores/SE, Santana de Sâo Francisco/SE and Propria/SE. According to BRASIL (2012), the forestry code legalizes the marginal strips of any natural perennial or intermittent watercourse, from the edge of the channel of the regular bed, as areas of permanent preservation, which must have preserved vegetation cover.

However, the deforestation of the riparian vegetation along the Sâo Francisco River and its tributaries is mainly due to the planting of crops on the banks of the river, a place considered appropriate by farmers, delimiting large tracts of bare soil, prone to erosion (NUNES and PINTO, 2007).

There are rice plantations in Brejo Grande/SE and this cereal is sent for processing by boat from the town near the Fishermen's Association to the city of Penedo/AL (Figure 8).

Figure 8: Profile of the use and occupation of the lower Sâo Francisco in Sergipe.

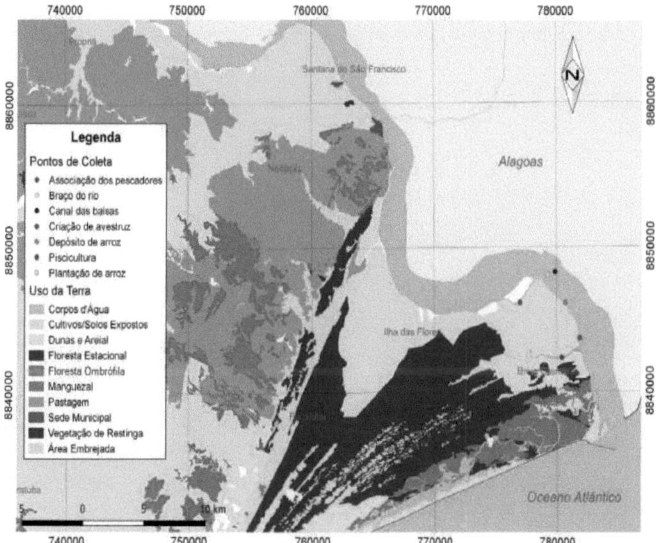

1.2.7 Characterization of the hydrological regime

The data used to characterize the hydrological regime in the lower Sâo Francisco was obtained through the electronic citizen information service (e-SIC), which intermediated the request to the National Water Agency (ANA) and the Sâo Francisco River Basin Committee (CBHSF). The hydrological information was recorded at the fluviometric station located in Propria (SE). For the station below, the historical series refers to the years 1980 to 2016.

1.2.8 Plotting the collection points

Using a boat located in the municipality of Brejo Grande (SE) and with the help of a Garmin e-trex digital GPS, latitude and longitude information was recorded in UTM to map the water collection points in the area of the mouth of the River San Francisco. The aim of defining the sampling points along the mouth of the River San Francisco was to assess the extent of the salt wedge and its quantification. In cross-sections from one end of the mouth to the other, considering a distance of approximately 150 meters.

At first, we sailed in the direction of the old Cabeço village, to mark out 13 points in cross-sections in the Foz area, between them, leaving the state of Sergipe in the direction of Alagoas (Figure 10). A further 32 points were plotted in the direction from the estuary to the municipality of Brejo Grande (SE) and the municipality of Piaçabuçu (Figure 11). In the transversal section, a distance of 150 meters was adopted between the Points and in the marginal section, an interval of 1km (Figure 9 A and B). The demarcated points were used to draw up the sample location maps. However, only 14 points were monitored in the field due to travel logistics and financial resources. The points evaluated are highlighted in Table 1 and 2.

Figure 9 A and B: Demarcation of the points along the mouth of the River San Francisco.

Table 1: Coordinates of the points located in the municipality of Brejo

LOCATION	COORDINATE X and Y
01	785355 e 8838872
02	785133 e 8838703
03	784911 e 8838535
04	784689 e 8838366
05	783538 e 8838732
06	784040 e 8839037
07	784543 and 8839343 (Distance from Foz 1.65 km)
08	785043 e 8839654
09	784822 e 8840515
10	784166 e 8840478
11	783510 and 8840442 (Distance from Foz 3.05 km)
12	782854 e 8840405
13	782792 e 8841029

14	782939 and 8841062 (Distance from Foz 3.9 km)
15	782373 e 8841824
16	782520 e 8841854
17	782138 and 8842634 (Distance from Foz 5.68 km)
18	782283 e 8842671
19	780645 e 8842313
20	781951 e8843441
21	782095 e 8843482
22	781810 e 8844253
23	781947 e 8844313
24	781532 and 8845002 (Distance from Foz 8.00 km)
25	781674 e 8845052
26	781063 e 8845670
27	781199 and 8845733 (Distance from Foz 8.87 km)
28	780559 e 8846273
29	779884 and 8846702 (Distance from Foz 10.34 km)
30	779266 e 8847372
31	778525 e 8847857
32	777878 e 8847360
33	777409 e 8846713
34	776777 e 8846197
35	780098 e 8849206

Table 2: Coordinates of the points located in the municipality of

LOCATION	COORDINATEDXeY
36	785147 e 8838165
37	785115 e 8839277
38	784894 and 8840327 (Distance from Foz 2.52km)
39	784368 and 8841237 (Distance from Foz 3.47 km)
40	783786 and 8842150 (Distance from Foz 4.9 km)
41	783951 e 8842338
42	782440 e 8841484
43	783261 and 8843111 (Distance from Foz 5.56 km)
44	782692 and 8844041 (Distance from Foz 6.64 km)
45	782762 e 8844453
46	781722 e 8844122
47	782017 e 8844834
48	781608 and 8845946 (Distance from Foz 8.82 km)

Figure 10: Transverse and longitudinal points on the Sergipe bank of the Foz do Rio Sao Francisco.

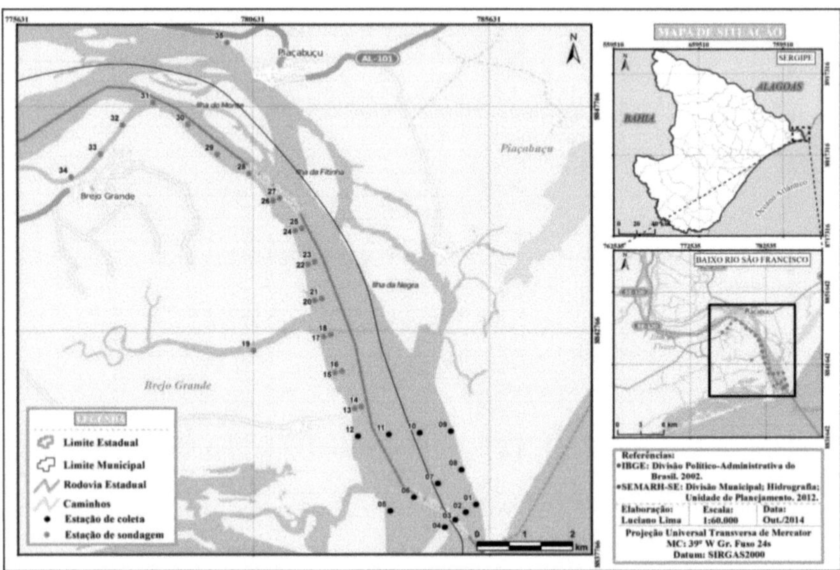

Figure 11: Transverse and longitudinal points on the Alagoas bank of the Foz do Rio Sao Francisco.

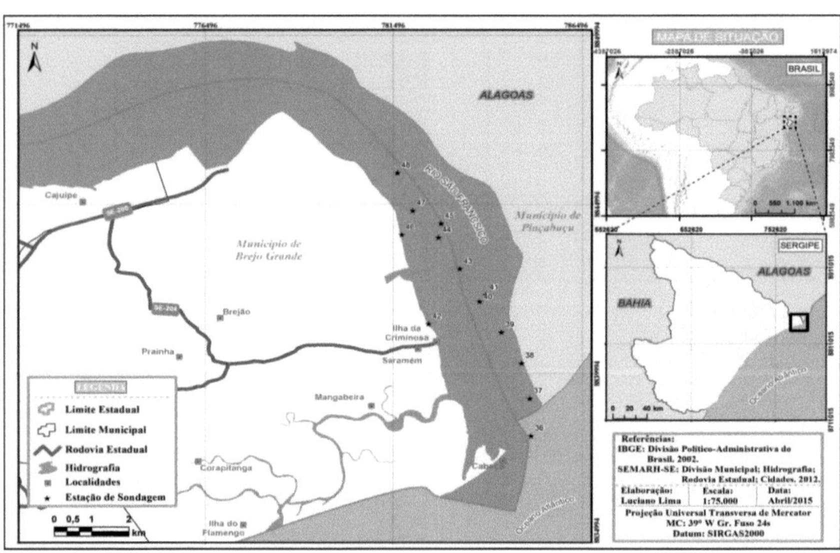

1.2.9 Characterization of the estuarine profile

1.2.10   Using the Acoustic Doppler Profiling (ADCP) method, a bathymetry section was carried out at the mouth of the River San Francisco to determine flow and depth. The ADCP is an operating system based on the Doppler effect, which measures flow through the apparent change in the frequency of waves reflected by suspended material (Doppler effect) (PITON, 2007). A limit was established for the bathymetry from the municipalities of Brejo Grande (SE) and Piaçabuçu (AI).

The river velocity and flow measurement campaigns at the mouth of the River San Francisco were carried out in two stages. The first measurement took place on March 15, 2015 and the second on March 8, 2016. The flow measurements were made using the Workhorse Rio Grande 600kHz Acoustic Doppler Current Profiler (ADCP) (Figure 12). The sections measured with the ADCP were obtained using the *WinRiver* II program for data processing and storage, which is capable of measuring current velocity, liquid discharge and section geometry.

During data acquisition, tests were carried out with different ADCP modes, which vary in the size of the measuring cells to measure at the shallowest points (banks) and also for locations with high current flow (Figure 13 A and B).

The survey of velocity and flow information aims to complement the studies on the physico-chemical analysis of water quality and the intrusion of the salt wedge at the mouth of the River San Francisco.

The measurement sites were demarcated in order to obtain measurement data in the region of the mouth of the River San Francisco. The measurement profiles were defined in such a way as to characterize the speed and flow along the river. Due to the influence of the tide in this region, two measurements were taken in each profile to detail the geometry of the river cross-section during the flood and ebb tide cycles.

Figura 12:   Use of the ADCP in the Foz do Rio Sao Francisco area.

Figura 13:   Cross-sectional profile of the Sâo Francisco river mouth channel between the municipality of Brejo Grande and Piaçabuçu/AL. **AeB:** Cross-section of the main channel at the mouth of the River Sâo Francisco

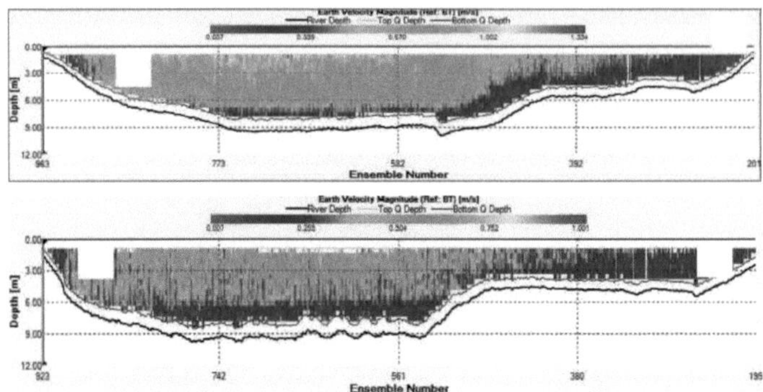

2.2.10   sample collection and measurement with Multiparametrcs probe.

The estuarine environment is governed by the influence of tidal oscillations. According to BRASIL (2016)[4] a tide is the vertical oscillation of the surface of the sea or other large body of water above the Earth, caused by the gravitational difference of the moon. Tides can be spring tides (or high tides) and ebb tides (or low tides).

The water samples were collected and measured with the multi-parameter probe during the spring tide (higher amplitude) and the ebb tide (lower amplitude). A total of eight campaigns were carried out, of which five were in the sizigia tide regime and three in quadratura. It should be noted that the sizigia tide was used as a reference, as this is the time of greatest scale of the sea's action on the river.

The time of collection was guided by the data contained in the tide table of the Port Authority of Sergipe (DHN, 2015), which contained the start and end times of the tidal cycles. The start of the tide was taken as the base for collection and always in a south to north direction, sailing upstream. The months evaluated included February, March, April, May, September and November in 2015, and March in 2016 (Table 2). In the month of September, the collection took place during an 11-hour tidal cycle, adopting the values from that time for salinity validation (information on the tidal cycle see Table 3).

Table 2: Information on the dates, type of tide and period used at the mouth of the Sâo Francisco River.

CAMPAIGN	DATE	TIDE	PERIOD
1	13.02.2015	Quadrature	Dry
2	05.03.2015	Sizigia	Dry
3	14.04.2015	Quadrature	Dry
4	20.05.2015	Sizigia	Rainy
5	26.09.2015	Sizigia	Rainy
6	28.11.2015	Sizigia	Dry

---

3 https://www.mar.mil.br/dhn/bhmn/download/cap10.pdf

| 7 | 02.03.2016 | Quadrature | Dry |
| 8 | 08.03.2016 | Sizigia | Dry |

Table 3: Sizigia tide forecast in the Foz do Sâo Francisco area, with a 24-hour cycle.

Day	Time	Height (m)
Saturday	08:53	0.0
Saturday	14:58	2.2
Day	Time	Height (m)
Sunday	09:36	-0.1
Sunday	15:41	2.3

Source: (DHN, 2015).

Salinity was measured using a HANNA HI9828 multi-parameter probe (Figure 14) at the collection points. The probe recorded salinity in (PSU) and transformed it into ‰. Electrical conductivity, turbidity, dissolved oxygen, total solids, temperature and pH were also checked. For the other parameters, the means of determination are shown in Table 4. Soil samples were collected using a "van Veen" type dredger (Figure 15) at two points in the main channel of the estuary. All these samples were packed in plastic bags and stored in thermal boxes, then sent to ITPS (Sergipe State Research and Technology Institute) for determination of granulometry and organic matter content.

Figures 14: Boat with multiparameter probe and equipment for collecting samples;

Figure 15: Use of the Van Veen dredger to collect soil.

Source: Personal collection.

Water samples were collected from taps in the municipality of Brejo Grande (SE) and in the village of Saramém (Figures 16 and 17). The samples were collected at random, 15 of which belonged to the village of Saramém and the other 15 to the town of Brejo Grande. Collection took place at random, observing the main streets and avenues of each locality.

The samples taken from the taps in the establishments did not come from a storage tank; this precaution is part of the analytical procedure. The samples taken from the taps came directly from the public network, the first jets of water were discarded and then the water was collected using 500mL polyethylene bottles (Figure 17). The parameters assessed were: pH, sodium and chlorides.

Figures 16: Collection of water samples from homes.

Figures 17: Conditioning of the water samples.

Source: Personal collection.

Table 3: Analytical methodology used to determine the physical and chemical parameters of the water collected in the area of the mouth of the Sâo Francisco River.

Variable	Unit	Methodology*
Conductivity	(µS.cm-1)	HANNA Multiparameter Probe - HI9829
Chloride	(mg L-1)	SMEWW$_1$ 2U12
Dissolved Oxygen	(mg L-1)	HANNA Multiparameter Probe - HI9829
pH	-	HANNA Multiparameter Probe - HI9829
Salinity	(mg L-1) and	Analytical method/Multiparameter probe HANNA - HI9829
Sodium	"'(mg L-1)	SMEWWi 2U12ı 45UU-Cl B
Total dissolved solids-STD	(mg L-1)	Analytical method/Multiparameter probe HANNA - HI9829
Turbidity	FNU	HANNA Multiparameter Probe - HI9829
Organic matter	g/dm3	WB (colorimetric)
Granulometry - Sand (Boyoucos Hydrometer)	%	Bouyoucos densimeter
Granulometry - Clay (Boyoucos Hydrometer)	%	Bouyoucos densimeter
Granulometry - Silt (Boyoucos Hydrometer)	%	Bouyoucos densimeter
Textural Classification	%	Bouyoucos densimeter
Specification for soil type	-	MAP-IN n^0 U2 U9/1U/2U8

[x]Standard Methods 21.ed.AHHA,2UUb.

2.2.11 Socio-environmental Analysis

In qualitative research, the interview is used to collect descriptive data in the language of the subject. For Marconi and Lakatos (2008), it is a data collection procedure that helps with diagnosis and, according to Gil (2009), it is a form of social interaction with the research object.

A semi-structured interview form was therefore drawn up (APPENDIX), containing 15 questions, 2 of which were specific to salinity in the river course and in households. Trivinos (2009) explains that these questions do not arise by chance, they are the result of a theory that guides the

47

researcher and all the information that he or she has already gathered about the social phenomenon of interest, and that the semi-structured research process gives better results with different groups of people.

In order to gather socio-environmental data on the region of the mouth of the River San Francisco, visits were made to collect information on *site in* 2014, 2015 and 2016. Application of a questionnaire with a population interview script (in APPENDIX). This script had 18 questions formulated with the aim of verifying the characteristics of the population, their source of income, their use of water and the profile of those who live in or frequent the Foz do Sao Francisco. The questions were designed to verify the profile of the socio-environmental impact caused by the salinization of the waters in the region. The focus group technique was used to verify the presence of aquatic macrophytes along the mouth of the River San Francisco. The qualitative research will be based on the elaboration of focus groups, which Gomes and Barbosa (1999) conceptualize as an informal discussion group of reduced size, with the aim of obtaining in-depth qualitative information.

Lopes et al (2009) indicate that in this technique, the expressions of each individual who takes part in the dynamic are intervened upon by the other subjects, allowing data to be collected in a way that brings the subjects involved closer together and favors the exchange of information.

The formation of the focus groups led to a workshop in May 2015. A photo album with images of the aquatic vegetation (Figure 18A, B, C, D, E and F) developed on the banks of the river branch from the mouth to Brejo Grande (SE). These photographic images and slides were used as teaching material for the workshops.

Figure 18 (A, B, C, D, E and F): Species of macrophytes present in the lower San Francisco region (A: *Eichhornia; B: Salvinia; C: Montrichardia linifera; D. Egeria; E: Oxycaryum sp: Egeria; E: Oxycaryum sp.;* F: *Derbesia).*

2.2.12 Statistical analysis

The data obtained from the quantitative samples with the parameters evaluated will be submitted and processed using statistical techniques such as: mean, minimum, maximum, standard deviation and correlation. Graphs were drawn up using Sigmaplot 12.2 software. These procedures are fundamental to estimating the reliability of the results. For the qualitative data, PASWStatistics Base software (SPSS) was used. Descriptive statistical frequency analysis was carried out, using absolute values and percentages to show the results of the sample.

2.3 Results and discussion

2.3.2 Water quality at the mouth of the São Francisco River

The statistical descriptors calculated to check water quality at the mouth of the River San Francisco are described in Table 1. The pH data presented ranged from 6.80 to 9.47, with the highest value occurring in February 2015. The values show a tendency towards alkaline environments, as most of the pH values are above 7.0. Values of around 8.33 were recorded by Cotovicz Junior et al, (2016) in the estuary area of the River San Francisco, which could be related to the concentrations obtained in this study. The alkaline pH indicates that salinity can reach significant values in the system evaluated.

Paiva et al (2006) analyzed the waters of Guaraja Bay, where the pH data found varied between 5.5 and 7.0, justifying in their work that the salinity at the mouth of the estuary did not reach values higher than 5. This is not the case in the estuary of the River San Francisco, which can record salinity values of around 20‰ to 30‰ at its mouth (CHESF, 2011).

According to the values in Table 4 and 5, Point 17 had a pH value of 6.80, which was considered the lowest index of this parameter during the survey, while Point 14 had a pH of around 9.47, which was considered the highest value. Points 14 and 17 are located in the estuarine region on the shore of the municipality of Brejo Grande/SE, Point 14 is within the range of significant salinity interference, according to Cavalcanti, Miranda and Medeiros (2017) who report that at low tide saline waters penetrate up to 6.8km from the mouth area, which may be related to the maximum pH value found at Point 14 (Figure 19A).

The pH of most natural waters is in the range of 6 to 9, which is between the support conditions for all forms of life (DAVES and MARTEN 2016). Marota, dos Santos and Enrich-Prasto (2008) consider that pH is a determining factor in the composition of species in a given location, as it directly influences metabolic processes. Oliveira, Campos and Medeiros (2010) add that sudden changes in the pH of water can lead to the extinction of biota. In general, the values found are within the range established by Conama Resolution 357/2005, which stipulates that pH should be acceptable between 6 and 9 for the purposes of classifying a body of water. It is noteworthy that there was only one instance of a pH value above the permitted range, representing 1.78% of the values obtained and 98.22% (Figure 19B) within the acceptable range.

Table 4: Descriptive statistics for water pH variables at the mouth of the São Francisco River in Sergipe. Months of February, March, April, May, November 2015 and March 2016. ____________

		Analysis points						
Parameters		**P1**	**P38**	**P39**	**P40**	**P43**	**P44**	**P48**
	Average	7,75	7,76	7,55	7,50	7,52	7,23	7,59
PH	Min	7,64	7,58	7,47	7,26	6,97	7,18	6,80
	Max	7,84	7,89	7,79	7,89	7,89	7,69	7,69
	DP	0,09	0,16	0,28	0,41	0,37	0,47	0,10
	CV%	1,18	2,01	3,65	5,40	4,97	6,53	1,33

cv%(coeficiente of variance); SD(Standard Deviation).

Table 5: Descriptive statistics for water pH variables at the mouth of the São Francisco River in Sergipe. Months of February, March, April, May, November 2015 and March 2016. __________

		Analysis points						
Parameter		**P7**	**P11**	**P14**	**P24**	**P27**	**P29**	**P17**
	Average	8,19	8,11	8,09	7,85	7,73	7,75	7,67
pH	Min	7,64	7,58	7,47	7,26	6,97	7,18	6,80
	Max	8,66	8,54	9,47	8,52	8,26	8,40	8,26
	DP	0,42	0,41	0,64	0,47	0,46	0,39	0,56
	CV%	5,19	5,12	7,85	6,01	5,95	5,09	7,27

cv%(coefficient of variance); SD(Standard Deviation)

Figure 19: Graph of the maximum, average and minimum values for pH at the mouth of the São Francisco River during the sundial and quadrature periods (A) and the percentage of acceptability according to CONAMA357/2005 (B).

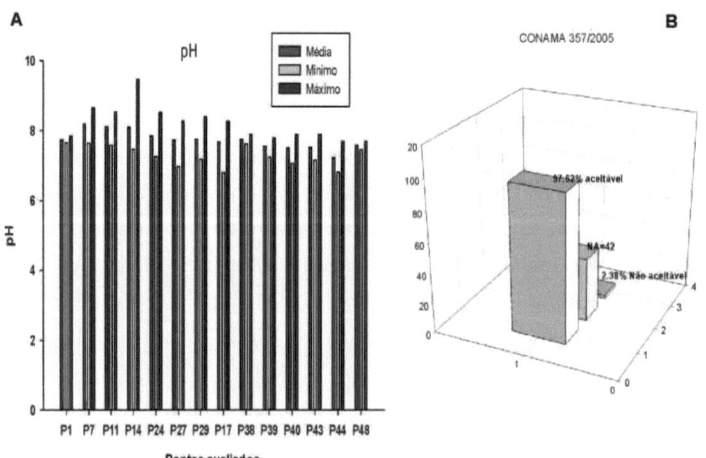

Dissolved oxygen is one of the main parameters for characterizing the effects of pollution resulting from discharges into a watercourse (OLIVEIRA, CAMPOS and MEDEIROS, 2010). Gonzaga (2014) adds that it is one of the main parameters for measuring water quality and is important for verifying the sanitary condition and maintenance of aquatic life and when recorded at low values indicates that the aquatic ecosystem is polluted. Jauzen, Shulz and Lamnon (2008) point out that the concentration of dissolved oxygen in water is the result of the interaction of various processes that tend to increase or decrease it, so the amount of dissolved oxygen is considered a primary indicator of water quality.

Dissolved oxygen values in the area evaluated ranged from 4.68 to 9.27 ppm, as shown in Tables 6 and 7. The lowest value was recorded at Point 14 in May 2015 and the highest at Point 17 in February 2015. CHESF - Companhia Hidroelétrica do São Francisco (2011) also recorded dissolved oxygen concentrations of 8.85 to 9.22 in February 2015, which may indicate that these areas are strongly aerated by the force of the tide or that there are no solid waste disposal points or significant organic discharges in the mouth region. It should be noted that all the municipalities around the mouth of the River San Francisco do not treat their effluents and in turn discharge their effluents into the river.

Franklin-Silva and Shettini (2003) explain that as a result of effluent discharges into the river course, the affected areas suffer large fluctuations in the dissolved oxygen content, harming the biota in general.

Fiorucci and Benedetti Filho (2005) report that it is common for waters polluted with organic substances associated with sewage to have a high demand for oxygen consumption, unless the water is continuously aerated. This indicates that the concentration of dissolved oxygen in water is the result of the interaction of various processes that tend to increase or decrease its concentration in aquatic environments (JANZEN, SCHULZ and LAMON, 2008). The average calculated for dissolved oxygen for the points analyzed makes it possible to classify this watercourse in Class 1, where the minimum value required to classify freshwater bodies of water in Classes 1, 2, 3 and 4 are, according to CONAMA Resolution 357/2005, 6 mg.L^{-1} , 5 mg.L^{-1} , 4 mg.L^{-1} , and 2 mg.L^{-1} (BUZELLI and SANTINO, 2012)(Figure 20A).

It is worth noting that values below 2 mg.L^{-1} for the concentration of dissolved oxygen in water can affect the health of the aquatic ecosystem and prevent the use of water for different purposes, including human supply (JANZEN, SCHULZ and LAMON, 2008). The oxygen values are shown in figure 20B, with 84% of the values obtained within the acceptable concentration according to CONAMA Resolution 357/2005 and only 16% of the results outside the permitted limit for class 2.

Table 6: Descriptive statistics for the oxygen variable in the water at the mouth of the São Francisco River in the Alagoas column. Months of February, March, April, May, November 2015 emar Ç° de2016 -

Parameter		TT	TOS-	POO-	PiO˙	∎Pi3-	Pii˙	P4Y
				Analysis points				
Dissolved Oxygen	Average	6,57	6,56	6,96	7,26	6,86	6,92	6,73
	Min	5,60	5,24	6,36	6,81	6,25	6,54	5,29

(mg.L⁻¹)	Max	6,98	7,01	7,64	8,19	7,59	7,30	7,30
	DP	0,66	0,88	0,53	0,64	0,69	0,39	0,96
	CV%	10,00	13,44	7,69	8,86	10,02	5,57	14,32

cv%(coefficient of variance); SD(Hadrao deviationJ.

Table 7: Descriptive statistics for the oxygen variable at the mouth of the São Francisco River, Sergipe column. Months of February, March, April, May, November 2015 emar o de2016 ç – ----

Parameter		Analysis points						
		P7⁻	W	"Pii""	▪Ta˙	T77⁻	▪Ta˙	TT7⁻
	Average	5,59	6,18	6,61	6,94	7,33	7,30	7,31
Dissolved Oxygen	Min	1,82	2,78	4,68	6,25	6,35	6,25	6,38
(mg.L⁻¹)	Max	7,33	7,23	8,99	8,76	9,27	9,65	9,27
	DP	2,00	1,46	1,29	0,78	1,11	1,42	1,12
	CV%	35,80	23,58	19,47	11,24	15,13	19,52	15,35

cv%(coeтiδienтe de vananδia)! SD(Standard deviation).

Figure 20: Graphs with descriptive statistics for the dissolved oxygen variable in the water at the mouth of the River San Francisco. Months of February, March, April, May, November 2015 and March 2016 (A) and percentage of acceptability according to CONAMA357/2005.

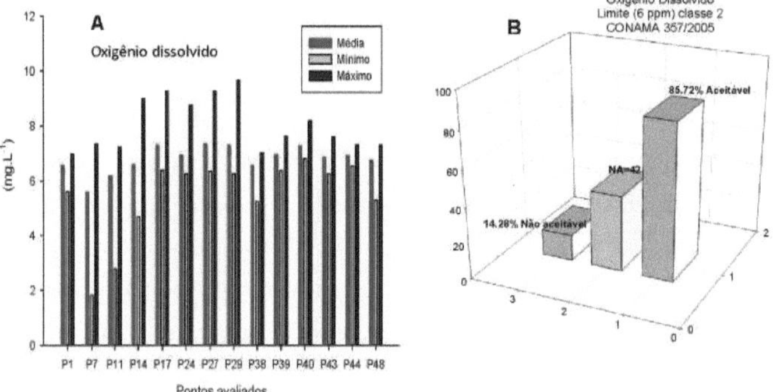

Total solids are also considered to be one of the parameters related to electrical conductivity and turbidity. Gonzaga (2013) reports that total dissolved solids are directly proportional to electrical conductivity. As total dissolved solids include salts, dissolved organic materials and nutrients, it can be said that very high concentrations of dissolved solids can increase both turbidity and electrical conductivity (VASCONCELOS, TUNDISI and TUNDISI 2009). Vasconcelos, Tundisi and Tundisi (2009) also point out that high concentrations of total dissolved solids can limit growth and lead to the death of many forms of life.

The total dissolved solids values ranged from 7.0 mg.L⁻¹ to 7,522 mg.L⁻¹ , as shown in Table 8. The highest concentration of total dissolved solids recorded occurred at Point 1 in the month of November, a dry period with a tidal regime of sizigia and a water level rise of 2.2m. Total solids values above 500 mg.L⁻¹ were also recorded by Saraiva et al (2009) in their research on the Lavagem stream, upstream of its mouth on the Espirito Santo river. According to Figure 21A and Table 8, only Point 43 showed values within those specified by CONAMA Resolution 357/2005.

52

Turbidity refers to the amount of suspended particles, which can hinder the transmission of light, altering life in the aquatic environment. It refers to the amount of suspended particles in the water. It is inversely proportional to the availability of solar radiation, which is essential for the primary production of an ecosystem (ALVES et al, 2012). The turbidity values recorded were between 2.9 NFU and 231.00 NFU (Table 7 and 8), in accordance with CONAMA Resolution 357/2005, which establishes a turbidity limit of up to 40FNU for class 1 and 100FNU for class 2; only Point 7 is within the limit for class 2. At Point 27, turbidity was recorded at 231 NFU during the dry season, a value above that proposed by the Resolution and limiting its use. The other points fall into class 1 (Figure 21A).

The highest values were recorded in March 2016, during a dry period and at high tide. Lima (2007), evaluating turbidity in the estuary of the Jacuipe River in Bahia, also recorded turbidity values of between 50 and 80 FNU during sizzling tides and during the dry season, coinciding with this study. Cunha et al (2005) adds that surface waters in estuarine rivers show a reduction in the levels of this variable in the rainy season, due to the increase in flow, as the rainfall supply favors water dilution and self-depuration.

Table 8: Descriptive statistics for the variables Total dissolved solids Turbidity in the water at the mouth of the Sâo Francisco River Alagoas column. Months of February, March, April, May, November 2015 and March 2016.

Points of analysis								
Parameter		P1	P38	P3У	P4U	P43	P44	P48
	Average	5603,75	4117,50	3226,25	1853,50	209,55	2298,00	2233,13
	Min	3356,00	2552,00	355,00	152,00	39,09	139,00	12,57
Solldos Totals Dlssolvldos	Max	7522,00	5004,00	6076,00	3468,00	383,00	4502,00	4502,00
(mg.L·)'	DP	1827,89	1075,82	3284,91	1810,45	191,76	2487,56	2562,46
	CV 7th°	32,62	26,13	101,82	97,68	91,51	108,25	114,75
	Average	9,35	8,88	11,23	7,90	12,50	3,68	10,73
Turbldez	Min	5,70	4,90	10,50	5,00	6,60	2,90	3,80
(FNU)	Max	14,30	17,00	12,40	11,40	24,20	4,80	22,90
	DP	3,85	5,64	0,88	3,38	8,30	0,84	8,79
	CV7	41,17	63,59	7,85	42,84	66,38	22,81	81,95

cv7o(coeтıcıenre ae variartela;; DP(Desvıo Hadrao).

Table 9 Descriptive statistics for the variables Total dissolved solids Turbidity in the water at the mouth of the Sâo Francisco River Sergipe column. Months of February, March, April, May, November 2015 and March 2016.

Points of analysis								
Parameter		PZ	P11	P14	P24	P2/	P2У	P1/
	Average	3122,14	2353,87	3067,42	125,67	1048,88	610,25	982,00
	Min	14,40	27,78	19,38	10,50	7,00	57,00	54,00
Total Dissolved Solids	Max	6566,00	5001,00	6289,00	441,00	2293,00	2552,00	2492,00

(mg.L)[1]	DP	2771,79	1701,11	1973,53	170,72	1003,44	876,99	1123,97
	CV%	88,78	72,27	64,34	135,85	95,67	143,71	114,46
	Average	31,76	13,11	12,63	11,58	36,15	9,03	6,58
Turbidity	Min	3,60	3,60	3,90	3,80	2,90	3,60	3,10
(FNU)	Max	99,10	29,30	38,20	38,40	231,00	25,90	17,00
	DP	39,23	9,50	11,60	11,18	78,91	7,53	5,18
	CV%	123,49	72,43	91,91	96,58	218,27	83,39	78,81

UV%(LoeTlclehTe de Vaɲarrclaj; DP(Deδ√l0 radrâô).

Figures 21: Graphs with descriptive statistics for the total dissolved solids variables at the mouth of the Sâo Francisco River. Months of February, March, April, May, November 2015 and March 2016 (A) and percentage of acceptability according to CONAMA357/2005 (B).

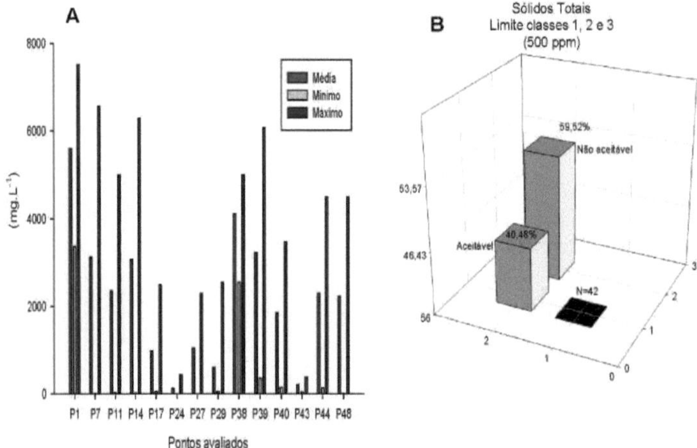

Figures 22: Graphs with descriptive statistics for the variables Water turbidity at the mouth of the River San Francisco. Months of February, March, April, May, November 2015 and March 2016 (A) and percentage of acceptability according to CONAMA357/2005 (B).

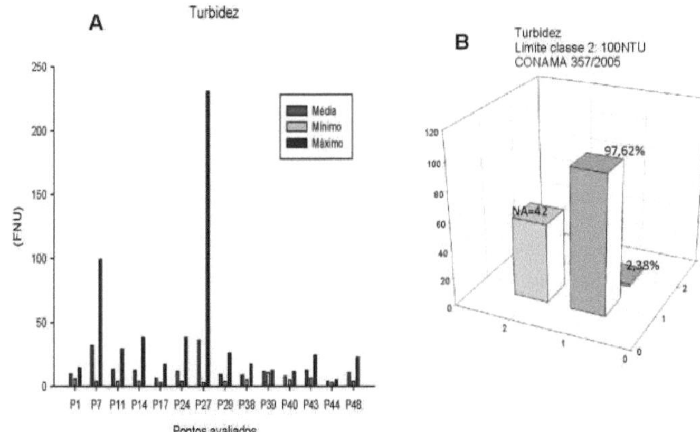

Electrical conductivity values ranged from 10.00 (μS/cm^{-1}) to 13130.00 (μS/cm^{-1}), with the highest value recorded at Point 7 and the lowest at Point 38. According to Esteves (1998), electrical conductivity is a measure of water's ability to conduct an electric current. Fritzsons; Mantovani, Chaves Neto and Hindi (2009) add that electrical conductivity, in turn, provides an indication of the salinity of a solution or, roughly speaking, the degree of mineralization of the water.

The highest value recorded was in May 2015, when the tide was on the ebb and flow and the rainy season, and it should be noted that Point 7 is at the entrance to the mouth of the River San Francisco. CETESB (2010) states that levels above 100 (μS/cm^{-1}) in freshwater indicate impacted environments. However, electrical conductivity in estuarine fluvial environments can be correlated with salinity. Baldotto and Canellas (2011) report that the influence of the sea is characterized by an increase in pH, ionic strength and electrical conductivity.

Paiva et al (2006) in their research found low conductivity and salinity values in periods of lower river discharge, at the end of the ebb and beginning of the flood, the waters coming from the Guama and Guajara-Açù rivers in the Para region have relatively low conductivity-salinity values (100-300(μS/cm^{-1}) - 0.04-0.12‰). On the other hand, in the so-called flood channel, adjacent to the shore where the city of Belém is located, conductivity-salinity values were between 350 and 1000 (μS/cm^{-1}); 0.15 to 0.42‰.

The salinity values for Point 7 showed maximum values of 38.03‰, correlating with the increase in electrical conductivity. The correlation between electrical conductivity and salinity is shown in Figure 13. Saraiva et al (2009) point out that although there are no standardized water quality standards for electrical conductivity, levels above 100 (μS/cm^{-1}) are generally adopted as a reference to indicate impacted freshwater environments.

Salinity is a dimensionless quantity of water that enables the characterization of bodies of water, the conditions for multiple uses to meet different demands (AGUADÊ, 2014).

In estuaries, salinity fluctuates both from place to place and from time to time, when seawater, with an average of 35‰, mixes with freshwater 0‰ between this range. Salinity varies according to depth, as salt water is denser and remains at the bottom (CASTRO and HUBER, 2012). Tides are the main mechanism for the penetration of saline waters. In places where there is little influence from fresh water through rivers or rain, salinity can be high, as the tides reach these places carrying water and salt (SIMÕES, 2007).

According to Segundo (2001), salinity distribution in the waters of the mouth of the São Francisco River is regulated by river flow and tidal action.

The average salinity concentration ranged from 0.63 to 14.07‰ (Table 11), with the highest values recorded at Points 7 and 11 located on the shore of the municipality of Brejo Grande/SE and Points 39 and 43 located towards Piaçabuçu in the state of Alagoas (Table 10). For the aforementioned points, salinity ranged between 38.03 and 38.95‰ in the direction of Sergipe and 35.00 and 37.00‰ in the direction of Alagoas. However, there were records of occasional maximum values between these points ranging from 37 to 65‰, these being in the layer below the surface, Castro and Huber (2012) indicate that salinity flows along the bottom in what is often known as a saline wedge. Freshwater, which is less dense, flows along the surface.

Loitzenbauer and Mendes (2011) add that the process of continental freshwater discharge meeting the action of waves and tides with salt water generates a salinity gradient from the continent towards the ocean. If continental discharge decreases, marine action is accentuated, increasing the estuarine area or the concentration of salts in the inner region of the estuary. If continental discharge increases, the mixture moves towards the ocean.   towards the ocean.

According to Medeiros et al (2014), the estuary of the São Francisco River is a salt wedge, with the tributary flows in the river determining the intensity of salinity in the estuary, as well as its extent upstream of the mouth, with the flow being the main human-controlled factor that determines the magnitude and extent of the salt wedge in the estuary.

The most significant salinity concentrations were recorded in March 2015 and May 2015, both during sizigia and flood tides. The information recorded is in agreement with Cavalcanti, Miranda and Medeiros (2017), who found that the action intensified during the period of sizigia. The authors found that salinity showed strong vertical stratification, varying from the surface to the bottom between 0.0 to 36.6 ‰ and 0.5 to 36.1‰, during syzygy and quadrature, respectively. Schettini (2002) indicates that tropical waters can have a salinity of more than 35‰;

Pereira et al (2010) analyzed the tidally dominated Caravelas/BA estuary and found salinity values between 34 and 36.5‰o. Segundo (2001), in an analysis of the estuary of the River San Francisco, recorded surface salinity ranging from 0.1 to 18.9‰ and bottom salinity ranging from 9.2 to 32.2‰. According to the author, the maximum peak at the bottom occurred after the salt wedge entered the estuary. This is a similar scenario to our research, as the highest salinity records were obtained after the entrance of the salt wedge.

The average salinity values according to CONAMA Resolution 357/2005 make it possible to classify 92.83% of the estuary as brackish water, since the data in the concentration range 0.5‰ and less than 30‰; according to the resolution (Figure 23).

The São Francisco River has faced dams along the main channel for several decades and after the construction of the Xingô HPP, the river underwent a new regularization, putting an end to flooding in the region (OLIVEIRA et al. 2001). The flow regime determined by the regularization of the river for energy generation purposes is a potential source of conflict between public supply, irrigation and environmental flows (MARTINS et al 2011).

Table 10: Absolute statistics for the variables electrical conductivity, salinity and temperature for the water at the mouth of the Sao Francisco river, Alagoas column. Months of February, March, April, March, November 2015 and March 2016.

Analysis points								
Parameter		**P1**	**P38**	**P8У**	**P40**	**P48**	**P44**	**P48**
	Méala	6609,2/	5928,00	5030,00	3671,25	3718,75	2298,00	2233,13
Electrical Conductivity	Mln	13,09	10,01	725,00	312,00	737,00	139,00	12,57
(µS/cm)'	Max	9985,00	8906,00	9384,00	6797,00	6732,00	4502,00	4502,00
	DP	4693,26	4170,44	4970,00	3559,34	3421,94	2487,56	2562,46
	cv%	71,01	70,35	98,81	96,95	92,02	108,25	114,75
	Méala	6,33	4,56	21,37	10,58	20,39	1,12	3,16
Sallnidade	Mln	3,49	2,75	7,73	3,66	4,27	0,06	1,12
(‰)	Max	8,73	5,61	35,00	20,00	37,00	2,68	7,17
	DP	2,29	1,25	15,18	7,82	18,61	1,29	2,75
	cv%	36,23	27,35	71,04	73,89	91,30	114,70	86,91
	Méala	28,98	28,93	28,91	29,13	29,14	29,24	29,18
Temperature	Mln	28,60	28,35	28,75	28,73	28,78	28,79	28,52
(^{0} C)	Max	29,23	29,23	29,03	29,46	29,49	29,67	29,67
	DP	0,31	0,42	0,14	0,39	0,40	0,50	0,58
	cv%	1,07	1,44	0,49	1,34	1,38	1,71	2,00

cv%(coetlciente aevanancia); DP(Desvlo Paarao).

Table 11: Aescrltlva statistic for the variables conautlv| aaae electrlc, Sallniaaae and temperature for the water at the mouth of the Rio São Franclsco column in Serglpe. Months of February, March, April, March, November 2015 and March 2016.

Hontos is analyzed								
Harameter		**H/**	**H11**	**H14**	**H24**	**H2/**	**H2У**	**H1/**
	M6dia	4318,24	4756,58	5156,97	1350,73	1573,03	1222,50	1417,03
Electronic Communication	Min	10,73	55,38	40,74	17,37	13,00	113,00	46,26
(µS/cm)'	Max	13130,00	10000,00	12570,00	5440,00	4509,00	5268,00	4954,00
	DP	5030,42	3422,22	3896,73	2008,77	1905,37	1792,75	2091,21
	CV 7th°	116,49	71,95	75,56	148,72	121,13	146,65	147,58
	M6dia	14,07	12,99	6,03	6,57	1,10	0,63	1,03
Sallniaaae	Min	3,20	2,02	1,18	0,06	0,00	0,05	0,05
(‰)	Max	38,03	38,95	24,31	33,17	2,38	2,74	2,67

	DP	11,33	15,44	7,61	11,62	1,07	0,95	1,19
	CV 7th°	80,55	118,87	126,22	176,94	96,58	149,89	115,79
	M6dia	28,75	28,74	28,85	28,80	29,68	28,79	28,74
Temperature	Min	26,03	25,90	26,20	26,08	26,08	26,18	26,08
(^{0} C)	Max	30,50	30,49	30,50	30,30	30,30	30,30	30,30
	DP	1,68	1,90	1,78	1,72	10,40	1,71	1,72
	CV 7th°	5,85	6,61	6,15	5,97	35,06	5,95	5,97

CV7o(uoeTicieNie ae vanancia); DP(Desvio Hadrao).

Figures 23: Average and maximum values for salinity at the mouth of the Sâo Francisco River during the syzygy and quadrature periods (A). Percentage of salinity in the data obtained (B).

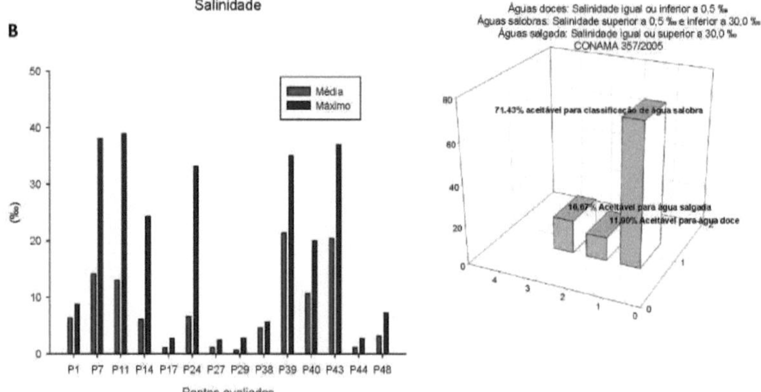

2.3.2 Quality of drinking water abstracted from the mouth of the Sâo Francisco River

The results obtained for pH in drinking water between the municipalities of Brejo Grande/SE and the village of Saramém showed average values within the limit established by Ministry of Health Ordinance 2914/2011. The aforementioned Ordinance allows for a pH range between 6.0 and 9.5. It should be noted that, among the minimum values, there were records of pH 5.74 to 5.86 between May, August, September and October 2016, months that are already considered rainy, which may be responsible for the decrease in pH. It should be noted that among the maximum values presented, the pH ranged from 8.38 to 8.57, indicating an alkaline characteristic (Table 12 and Figure 23A).

Magalhâes et al (2014) indicate that the pH parameter in drinking water should be frequently assessed, as it is an indicator that can potentiate toxic elements or bacterial media contained in the water. Thus, control at the time of collection and distribution must be monitored with treatment techniques that allow water to be consumed without risking human health. Carmo, Bevilacqua and Bastos (2008) state that the quality of the water consumed results from the quality of the raw water and the rigorous operational control of the treatment processes, i.e. from the point where the water is abstracted from the source to the point of consumption.

Water for human consumption is one of the most important carriers of disease, which makes its evaluation essential. Scalize et al (2014) found values with a similar proportion to those presented in this study, with a single representation of pH below 6.0, the others being in the range recommended by legislation, which is 6.0 to 9.5.

Table 12: Descriptive statistics for the pH variable of drinking water quality in the municipality of Brejo GrandeZSE and Povoado Saramém.

PARAMETERS	24.10.14	15.01.15	05.03.15	20.03.15	21.05.15	27.08.15	27.09.15	28.10.15
pH Limit MS.2914Z2011 (6,0 a 9,5.)	Average 6.65	6,65	6,79	6,87	6,60	6,29	6,59	6,57
	Minimum 6,52	6,52	6,20	6,29	5,86	5,77	5,74	5,74
	Maximum6,91	6,91	7,30	8,52	8,57	8,38	7,92	7,92

In figure 24(B), the proportional representation of the samples with values in line with the Ministry of Health's legislation indicated 83.4% within the established limit and only 16.67% below the legislated standard value.

Figures 24: Graph of the descriptive statistics for the pH variable for drinking water quality in the municipality of Brejo GrandeZSE and Povoado Saramém (A) and (B) acceptability according to the Ministry of Health Ordinance.

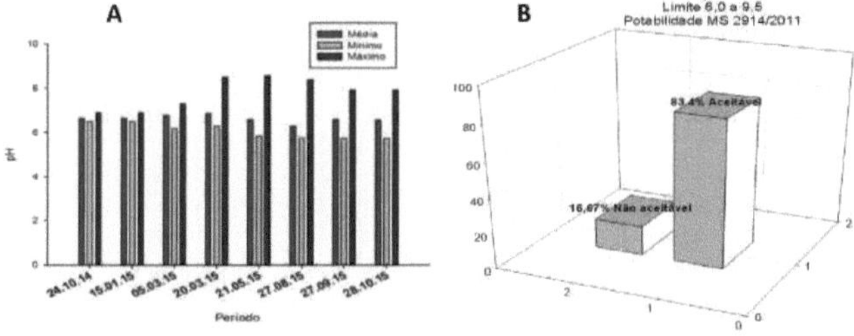

When assessing the sodium and chloride parameters, the average values obtained in May 2015 ranged from 1962.85 mg.L^{-1} to 2938.62 mg.L^{-1} (Table 12). These values showed a high concentration of sodium and chlorides present in drinking water and above the limit set by MS Ordinance 2914/2011, which establishes a tolerance of up to 200 mg.L^{-1} for the sodium parameter and 250 mg.L^{-1} for chloride. It should be noted that among the maximum values recorded were concentrations of 6901.99 mg.L^{-1} sodium and 15010.50 mg.L^{-1} for chloride. However, the average values in the other months ranged from 6.03 to 119.12 mg.L^{-1} for sodium and from 15.26 to 247.22 mg.L^{-1} for chloride (Table 13 and Figure 25). Sodium and chloride can change the taste of water, making it unsuitable for consumption and consequently affecting human health. Morgano (2002) states that levels of sodium above 20 mg.L^{-1} in both drinking water and water used in food preparation can already be considered a source of risk for individuals suffering from kidney problems and hypertension.

The municipality of Brejo Grande/SE and the village of Saramém are located in the area of the mouth of the River San Francisco, environments influenced by tidal oscillations. It is worth noting that the raw water collection point for the village of Saramém is located on the Paraùna River, which is fed by the tidal current, with the possible ingress of sodium and chloride elements from marine waters. The raw water collection point for the population of Brejo Grande/SE is located on the main course of the River San Francisco near the fishermen's headquarters in the region.

According to Cunha (2005), a large part of the world's population lives in urban centers and in areas close to estuarine coastlines, making this a strong component of the impact on natural bodies of water. In response to the semi-structured interviews, residents reported that the change in the taste of the water was perceived during high tide (Table 14).

Frank et al, (2012) add that the taste produced by chloride ion varies with its concentration and also depending on the chemical composition of the water. Thus, water with a concentration of 250mg/l can have a noticeable saline taste when the cation present in the solution is sodium (Na+).

Medeiros, Lima and Guimaraes (2016) recorded chloride values within the limit established by Ordinance MS 2914/2011 in drinking water in riverside communities in the state of Para. This was partially similar to the average values in this study, except for the month of March 2015.

Figure 26 shows a correlation of 0.99 between the Sodium and Chloride parameters, indicating a mutual relationship between the two chemical elements.

Figures 25: Graphs of the average values of the Sodium (Na+) and Chloride (Cl-) parameters (A).

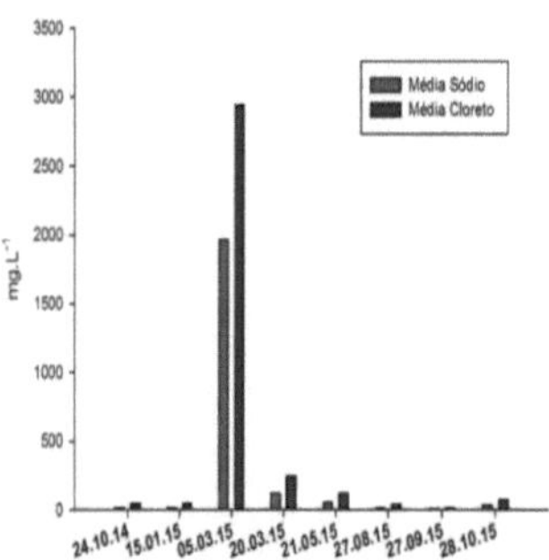

Figures 26: Pearson correlation graphs for drinking water in the municipalities of Brejo Grande/SE and Saramém(B).

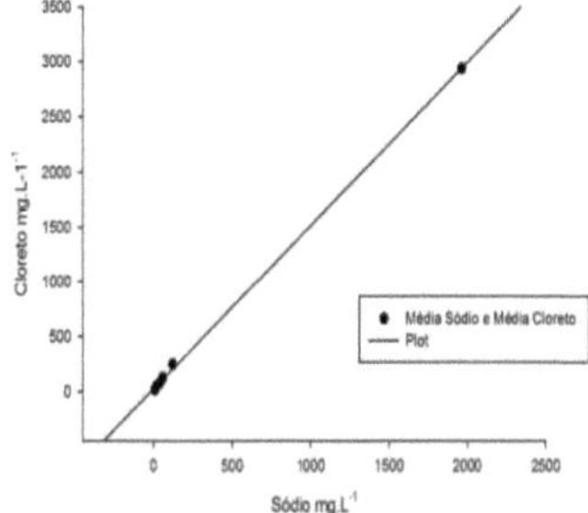

Source: Author (2016)

Table 13: Descriptive statistics for the variables Sodium (Na+) and Chloride (Cl-) for drinking water quality in the municipality of Brejo Grande/SE and Povoado Saramém.

PAPAMEimS			15.U1.15	U5.U3.15	2U.U3.15	21.U5.15	22.UU.15	22.UU.15	2U.1U.15
Sodium	Average	18,11	18,11	1962,85	119,12	56,31	16,19	6,03	37,93
Limit MS.2914/2011	Minimum	5,81	5,81	6,39	4,26	5,87	5,20	5,96	31,17
200 mg.L-1	Maximo	39,51	39,51	6901,99	456,04	180,81	35,33	6,16	42,49
Chloride	Average	47,40	47,40	2938,62	247,22	124,00	42,39	15,26	75,20
Limit MS.2914/2011									
250	Minimum	14,53	14,53	16,46	6,08	12,99	12,99	15,08	69,04
ı-ı mg.L	Maximo	100,72	100,72	15010,50	954,04	413,74	90,07	15,64	80,27

Table 14: Qualitative statistical evaluation of changes in the taste of drinking water in Povoado Saramém and Brejo Grande/SE.

Influence of the sea on the taste of water

ANSWERS	FREQUENCY	PERCENTAGE
High Tide	b2	96,36
Low Tide	3	6,64

2.4 Conclusion

At the mouth of the Sâo Francisco River, pH was found to be alkaline at most of the points assessed, conductivity and total dissolved solids were significant, but not indicative of sea salts, because turbidity did not show high concentrations of sediment or particles, elements that favour increased turbidity in the water.

The salinity concentrations were significant and indicated that the tide period was acting to increase salinity. It is noteworthy that the points with salinity above 16 (‰) were located in the main channels on the banks of the municipalities of Brejo Grande/SE and Piaçabuçu/AL.

Salinity may be associated with the regularization of flows, which has inhibited flood pulses and the force of the river on tidal movements towards the ocean. In addition to the advance of the sea, the mouth of the Sâo Francisco River faces specific degradations such as the dumping of solid waste and liquid effluents, disorderly occupation of its banks and the loss of riparian forest. All these problems accentuate the loss of water quality and compromise the sustainability of the aquatic ecosystem.

Water from the River San Francisco is used for a variety of purposes, including public supply, and should not pose a risk to the health of users. However, elements such as sodium and chloride have been recorded at levels above those permitted by the Ministry of Health's supervisory body, indicating that the effects of the tides could also compromise the quality of consumption for the riverside population. Although it was not the focus of this study, it is hypothesized that areas irrigated with water taken from the mouth of the River San Francisco may contain levels of sodium, which must be carried in the water used for crops.

Salinity is a worrying factor because it directly compromises water use and affects the region's biota. It is suggested that further research be carried out in the region in order to monitor the saline wedge and its potential risks for the population of the lower São Francisco.

REFERENCES

ALVES, I. C. C.; EL-ROBRINI, M.; SANTOS, M. L. S.; MONTEIRO, S. M.; BARBOSA, L. P. F.; GUIMARAES, J. T. F. Surface water quality and assessment of the trophic state of the Arari River (Marajó Island, northern Brazil). **Acta amazònica**. vol.42(1) 2012:115 -124

ANA. National Water Agency. Water resources plan for the São Francisco river basin. Brasilia. 2004. 337p.

ARAÙJO, S. S. Appropriation of natural resources and socio-environmental conflicts in the lower Sao Francisco in Sergipe and Alagoas. Thesis (PhD in Development and Environment) - Federal University of Sergipe, 2015. 663p.

BALDOTTO, M. A.; CANELLAS, L. P. Oxidation capacity as an index of sediment organic matter stability according to the fluvial-estuarine gradient of the Paraiba do Sul River. **Chem. Nova**, Vol. 34, No. 6, 973-978, 2011.

BRAZIL, CONAMA Resolution 357/2005. Classification of water bodies and guidelines for their classification and conditions and standards for effluent discharge. Available at: <http://www.mma.gov.br/port/CONAMA/>. Accessed on March 3, 2009.

BRAZIL. Law No. 12.651, of May 25, 2012. Provides for the protection of native vegetation; amends Laws Nos. 6.938, of August 31, 1981, 9.393, of December 19, 1996, and 11.428, of December 22, 2006; repeals Laws Nos. 4.771, of September 15, 1965, and 7.754, of April 14, 1989, and Provisional Measure No. 2.266-67, of August 24, 2001; and makes other provisions. Diario Oficial da Uniao, Brasilia, May 28, 2012.

BRAZIL. Ministry of Health. Provides for procedures to control and monitor the quality of water for human consumption and its standard of potability. Ordinance n^0 2914/MS of 2011. Available at:<

http://www.saude.mg.gov.br/images/documentos/PORTARIA%20No-

%202.914,%20DE%2012%20DE%20DEZEMBR0%20DE%202011.pdf> Accessed on: December 12, 2016.

BRITTO, F.B.; SILVA, T.M.M.; VASCO, A. N.; AGUIAR NETTO, A. O.; CARVALHO, C. M. Assessment of the risk of water contamination by pesticides in the betume irrigated perimeter on the lower São Francisco River. **Revista Brasileira de Agricultura Irrigada** v.9, no.3, p. 158- 170, 2015.

BUZELLI, G. M.; SANTINO, M. B. C. Analysis and diagnosis of water quality and trophic state of the Barra Bonita reservoir, SP. **Revista Ambiente & Agua** - An Interdisciplinary Journal of Applied Science: v. 8, n.1,2013.

CABRAL, S. A. S.; AZEVEDO JUNIOR, S. M.; LARRAZABAL, M. E. .Seasonal abundance of migratory birds in the Piaçabuçu Environmental Protection Area, Alagoas, Brazil. **Rev. Bras. Zool**. [online]. 2006, vol.23, n.3, pp.865869. ISSN 0101-8175.

CARMO, R. F.; BEVILACQUA, P. D.; BASTOS, R. K. X. Surveillance of drinking water quality: a qualitative approach to hazard identification. **Eng. sanit. ambient**. Vol.13 - N 4 - Oct/Dec 2008, 426-434

CARVALHO, M. E. S. and A. L. FONTES. Shrimp farming in Sergipe's coastal area. **Revista da FAPESE de Pesquisa e Extensao**. v. 3 n 1, p. 87-112. 2007.

CASTRO, PETER; HEBER, M. E. **Marine Biology**. 8. ed. Porto Alegre: AMGH, 2012. 461 p

CAVALCANTE, G.; MIRANDA, L. B.; MEDEIROS, P. R. P. Circulation and salt balance in the Sao Francisco river Estuary (NE/Brazil). **RBRH**, Porto Alegre, v. 22, e31,2017.

CBHSF-São Francisco River Basin Committee. Consolidated diagnosis of the São Francisco river basin 2016-2025. Volume 1. 489 p. 2016.

CBHSF-San Francisco River Basin Committee. Technical report of the campaign to assess the socio-environmental changes resulting from the regularization of flows in the lower São Francisco River. Maceiò, AL, 2013, 175p.

CHESF-Sao Francisco Hydroelectric Company. Inventory of the Aquatic Ecosystems of the Lower Sao Francisco River. 3^0 Annual Report. December/2009 to November/2010. 2011. 427p

COMPANHIA DE TECNOLOGIA DE SANEAMENTO AMBIENTAL - CETESB (2008) Qualidade das aguas interiores no Estado de Sao Paulo. Sao Paulo: CETESB. 41 p

COSTA, F. J. C. B. Conservation of Neartic Migratory Birds in Brazil Recomposition of rheophilic ichthyofauna. Project for the integrated management of activities carried out on the lands of the lower São Francisco. Final report Canindé de Sao Francisco. Xingó Institute for Scientific and Technological Development. 74p. 2003.

COTOVICZ JR. L. C.; LIBARDONIA, B. G.; BRANDINIA,N.; KNOPPERSA, B.

A. GWENAEL, A. Comparisons between real-time measurements of aquatic pco2 and indirect estimates in two contrasting tropical estuaries: the eutrophicated estuary of Guanabara Bay (RJ) and the oligotrophic estuary of the Sao Francisco River (AL). **Chem. Nova,** Vol. 39, No. 10, 1206-1214, 2016

COUTO, A. G. Analysis of the flow regime of the Pedra do Cavalo HPP on the spatial and temporal behavior of the salinity of the fluvial-estuarine stretch of the lower Paraguapu River to Iguape Bay. Dissertation. MAASA. UFBA. 2014. 143p.

CUNHA, A. C.; CUNHA, H. F.; SOUZA, J. A.; NAZARE, A.; PANTOJA, S. Monitoring of Surface Water in Estuarine Rivers of the State of Amapa under Microbiological Pollution. Boletim do Museu Paraense Emilio **Goeldi. Ciências Naturais,** v. 1,n.1, p. 191-199, 2005.

DHN. Department of Hydrography and Navigation. Tide forecasts (daily highs and lows). Available at: <http://www. mar. mil.br/dhn/chm/box-previsao-mare/tabuas/> accessed on: 01.02.2015.

ESTEVES, F. A. **Fundamentos de limnologia.** 3 ed. Editora Interciencia. Rio de Janeiro, 2011.

FERREIRA, R. A.; PRATA, A. P.; ARAGAO, A. G.; SILVA, A. C. C.; MACHADO, W. J. The vegetation of the lower São Francisco as a subsidy for the recovery of degraded areas. 91-112p. In: Aguiar Netto, A. O.; SANTANA, N. R. F. **Contexto Socioambiental** das aguas do rio Sao Francisco. UFS. 2015. 342p.

FREITAS, M. B.; BRILHANTE, O. M.; ALMEIDA, M.; Importance of water analysis for public health in two regions of the State of Rio de Janeiro: focus on fecal coliforms, nitrate and aluminum. **Cad. Saùde Pùblica,** Rio de Janeiro, 17(3):651-660, mai-jun, 2001.

FRITZSONS, E; MANTOVANI, L. E; CHAVES NETO, A; HINDI, E. C. A influência das actividades mineradoras na alteração do pH e da alcalinidade em água fluvial: o exemplo do rio Capivari, regiao do carste paranaense. **Eng. Sanit. Ambient.** 2009, vol.14, n.3, pp.381-390. ISSN 1413-4152.

JANSEN, J. G.; SCHULZ, H. E.; Measurements of dissolved oxygen concentration at the water surface. **Eng. sanit. Environ.** Vol.13 - N^0 3 - jul/set 2008, 278-283.

JESUS, N. B.; GOMES, L. J. Socio-environmental conflicts in the extraction of aroeira (Schinus terebebinthifolius Raddi), Baixo Sao Francisco - Sergipe/Alagoas. **Ambient. soc.** 2012, vol.15, n.3, pp.55-73.

LIMA, G. M. P. Sedimentological characterization and patterns of circulation and mixing in the estuary of the Jacuipe River - north coast of the state of Bahia. Dissertation Institute of Geosciences Postgraduate course in Geology. UFBA. 85p

LOITZENBAUER, E.; MENDES, C. A. B. Salinity dynamics as a tool for integrated water resources management in the coastal zone: an application to the Brazilian reality. **Revista da Gestao Costeira Integrada** 11(2):233-245 (2011).

MAROTTA, H.; DOS SANTOS, R. O.; ENRICH-PRAST, A. Limnological monitoring: An instrument for the conservation of water resources in urban environmental planning and management. **Ambiente & Sociedade**. Campinas, v. XI, n.1, p.67-69, jan-jun 2008.

MARTINS, D. M. F.; CHAGAS, R. M.; MELO NETO, J. O. M.; MELLO JUNIOR, A. V. Impacts of the construction of the Sobradinho hydroelectric plant on the flow regime in the Lower São Francisco River. **Revista Brasileira de Engenharia Agricola e Ambiental**. v.15, n.9, p.1054-1061, 2011.

MEDEIROS, A. C.; LIMA, M. O.; GUIMARAES, R. M.; Exposure to urban and industrial pollutants in the municipalities of Abaetetuba and Barcarena in the state of Para, Brazil. **Ciência & Saùde Coletiva**, 21(3):695-708, 2016.

MEDEIROS, P.RP.; SANTOS, M. M.; CAVACANTE, G. H.; DE SOUZA, W, F. L.; SILVA, W.F. Environmental characteristics of the Lower Sao Francisco (A/SE): Effects of dams on the transport of materials at the ocean-continent interface. **Geochimica Brasiliense**. V 28. 65-78.2014.

MMA. Ministry of the Environment. The national system of nature conservation units. 2011. 16p

MORGANO, M. A.; SCHATTI, A. C; ENRIQUES, H. A; MANTOVANI, D. M. B. Physico-chemical evaluation of mineral waters sold in the region of Campinas, SP. **Ciênc. Tecnol. Aliment**. 2002, vol.22, n.3, p.239-243.

MOURA, A.C.; ASSUMPÇÂO, R.A.B.; BISCHOFF, J.; Physical, chemical and microbiological monitoring of the water of the Cascavel River from 2003 to 2006. **Arq. Inst. Biol.**, Sao Paulo, v.76, n.1, p.17-22, jan./mar., 2009.

NUNES, F. P.; PINTO, M. T. C. Local knowledge on the importance of riparian reforestation for environmental conservation in the Upper Sao Francisco, Minas Gerais. **Biota Neotrop**. [online]. 2007, vol.7, n.3, pp.171-179. ISSN 1676-0611.

OLIVEIRA, C. N. O;CAMPOS, V. P.; MEDEIROS, Y. D. P. Evaluation and identification of important parameters for the quality of water bodies in the semi-arid region of Bahia. Case study: Salitre river basin. **Chem. Nova,** Vol. 33, No. 5, 1059-1066, 2010.

PAIVA, R. S.; ESKINAZI-LEÇA, E.; PASSAVANTE, J. Z. O.; CUNHA, M. G. G.

S.; MELO, N. F. A. C. Considerações ecològicas sobre o fitoplancton Considerações ecològicas sobre baia do Guajara e Foz do rio Guama, Para, Brasil ara, Brasil. Bol. **Mus. Ciências Naturais**, Belém, v. 1, n. 2, p. 133-146, May-Aug. 2006

PEREIRA, M. A. D.; SIEGLE, E.; MIRANDA, L. B.; SCHETTINI, C. A. F. Hydrodynamics and seasonal suspended particulate matter transport in a tidally dominated estuary: Caravelas Estuary (BA). **Revista Brasileira de Geofisica,** v28 n(3), p 427-444, 2010.

SANTOS, L. C. M. Estuarine lagoon system of the São Francisco River, coastal zone of Sergipe:

Land use and cover and environmental diagnosis of mangroves. Dissertation. Department of Postgraduate Studies in Environmental Sciences, University of São Paulo-USP. 2010. 130p

SARAIVA, V. K; NASCIMENTO, M. R. L.; PALMIERI, H. E. L.; JACOMINO, V. M. F. Evaluation of sediment quality - case study: Ribeirao Espirito Santo sub-basin, a tributary of the São Francisco River. **Quim. Nova [online]**. 2009, vol.32, n.8

SCALIZE, P. S.; BARROS, E. F. S.; SOARES, L. A.; HORA, K. E. R.; FERREIRA, N. C.; BAUMANN, L. R. F. Avaliação da qualidade da água para abastecimento no assentamento de reforma agraria Canudos, Estado de Goias. **Rev. Ambient. Agua** vol.9 no.4 Taubaté Oct./Dec. 2014

SCHETTINI, C. A. F. Caracterização Fisica do Estuario do Rio Itajai-açu, SC RBRH - **Revista Brasileira de Recursos Hidricos** Volume 7 n.1 Jan/Mar 2002, 123-142

SEGUNDO, G. H. C. Hydrodynamic sedimentological characterization of the estuary and delta of the São Francisco River. Dissertation. Department of Meteorology/CCEN/UFAL. 2001.125p.

SEMENSATTO JUNIOR, D.L. The estuarine system of the Sao Francisco delta - SE: environmental analysis based on the study of foraminifera and tecamebas. PhD thesis prepared for the Geosciences Postgraduate Program. Rio Claro (SP) 2006. 226p

SIGRIST, M. S.; CARVALHO, C. J. B. Detection of areas of endemism on two spatial scales using Parsimony Analysis of Endemicity (PAE): the Neotropical region and the Atlantic **Forest. Biota Neotrop.**, vol.8, n.4. 2008.

SILVA, M. G. Environmental modeling in the Poxim River basin - Agu/SE and its anthropic relations. PhD thesis. Department of development and environment. Federal University of Sergipe. 2013. 224p.

SIMOES, E. C. Environmental diagnosis of heavy metals and organochlorines in mangroves in the Baixada Santista and Cananéia/SP estuarine complexes. Dissertation. Institute of Chemistry, Federal University of Sergipe. USP. 2007. 183p

SOUZA, F. P.; PERTEL, M.; TEIXEIRA, T.; FERREIRA, A. V.; MENEZES, L. E. F.; PEREIRA, P. S. F.; Quality of the water supply of the tamarindo community in Campos dos Goytacazes/RJ. **Perpectiva on line**. Exact sciences engineering. 2015. 1-16.

VALENTE, R. M; SILVA, J. M. C.; STRAUBE, F. C.; NASCIMENTO, J. L. X. Conservation of neartic migratory birds in Brazil. Conservagao internacional, 2011 400 p.

VASCONCELOS, F. M.; TUNDISI, G. J.; MATSUMURA-TUNDISI, T. Avaliagao da qualidade de água. 1ª ed., **Sociedade Mineira de Engenheiros Agronomos,** 322 p., 2009.

CHAPTER 3: HYDRODYNAMICS OF THE SÃO FRANCISCO RIVER FLOOD WITH THE USE OF COMPUTATIONAL MODELING

SUMMARY

The hydrodynamics of the waters of a river flowing towards the sea establish an important control on the volume and direction of the watercourse. The study was carried out in the area of the mouth of the São Francisco River between the municipalities of Propria/SE and Piaçabuçu/AL. To calibrate the hydrodynamic model, the simulations were adjusted based on the parameters of a field campaign carried out on September 26 and 27, 2015, at the end of the quadrature tide and the beginning of the sizigia tide, when the vertically oscillating waters are at their highest, closing a 12-hour cycle. The hydrodynamic scenario showed the influence of the tide on the flow of the river in stretches above Piaçabuçu/AL, an occurrence that could lead to changes in environmental dynamics and alterations in water quality characteristics, to the detriment of controlling the flows released by the Xingó hydroelectric power station.   . ..

Keywords: Hydrodynamics, Flow, Marine current.

3.1   Introduction

The hydrodynamics of estuaries are a determining factor in the processes and dynamics of the local environment. The hydrodynamic oscillations that exist in estuary areas allow sea water to mix with river water, establishing processes of exchange and energy circulation in the aquatic ecosystem.

The dynamics of estuaries are basically governed by the actions of the tides and the flow of fresh water. For Shi and Lu (2011) this process of mixing between the waters of the coastal zone and the waters of the river enables the transportation of sediments and aquatic biogeochemistry. D'Aquino et al (2011a) add that estuaries are highly variable in terms of physical and biological properties, and that the estuarine hydrodynamic system is the response to changes in freshwater flow.

Together with the estuarine system, the mouth of the river is fundamental for the population and the balance of natural resources. However, currently, a large part of its marginal course and its river discharge are undergoing strong effects from anthropogenic actions, which are signaled by the advance of environmental degradation in these spaces. These alterations jeopardize the sustainability of large rivers, including the São Francisco river basin, which currently has a large number of its tributaries at risk.

One of the risk factors is the process of regularization of flows attributed to the construction of dams along the course of the San Francisco River (MARTINS et al 2011). According to Aragao,

Severiano and Moura (2015), the installation of dams causes changes to the integrity of rivers. Bortone, Ludwig and Xavier (2016) argue that their environmental and social impacts are unquestionable, however, the number of dams in the world already exceeds 800,000. Dams are defined as structures on a permanent or temporary watercourse (BRASIL, 2012). According to Derrosso and Ichikawa (2012) damming most of the time causes the flooding of productive land areas and the entire social and ecological system culminates together.

This has been observed in the construction of dams along the San Francisco River, which have been responsible for natural and social changes in this environment, causing problems that were not considered during the installation period. According to Medeiros et al (2007), one of the most notable changes caused by the construction of dams on rivers is the regularization of the flow, with the aim of supplying the water needed to generate electricity, causing a large reduction in the natural flow, causing an imbalance of energy between the river and the sea.

In this context, this study aims to analyze the hydrodynamic changes in the area of the mouth of the São Francisco River by means of hydrodynamic modeling.

3.2 Methodology

3.2.1 Study area

The São Francisco river basin has a drainage area of 639,219 km^2 , the main bed of which is 2,700 km long and has an average flow of 2,850 m^3 $.s^{-1}$, running through the states of Bahia, Pernambuco, Alagoas, Sergipe, Goias and the Federal District (CBHSF, 2015).

According to the National Water Agency (2005), this basin is divided into four physiographic regions: Upper, Middle, Middle and Lower Sao Francisco which, for planning purposes, these areas were subdivided into thirty-four small basins, and 12,821 micro-basins with the aim of delineating by stretches the main rivers in the region. The study was carried out in the Foz do Rio Sao Francisco area between the municipalities of Propria/SE and Piaçabuçu/AL (Figure 27), regions belonging to the lower Sao Francisco. The climate of the region is divided into tropical semi-arid and tropical semi-humid, evapotranspiration can vary from 600 to 700mm in the wet season and 750mm to 800mm in the dry season, annual rainfall can decrease from the coast, there are hydromorphic soils, including organic, gley and alluvial soils, which in their natural state are subject to periodic flooding, have limited fertility and are more suitable for growing rice (IPEA, 2002).

The coastal strip dives under the ocean and advances as the substrate of the continental edge and the rocks of the sedimentary basin constitute the geological substrate of the final stretch of the São Francisco river-sea system (FONTES, 2016).

Figure 27: Map of the São Francisco river basin and location of the study area between the municipalities of Própria/SE and Piaçabuçu/AL.

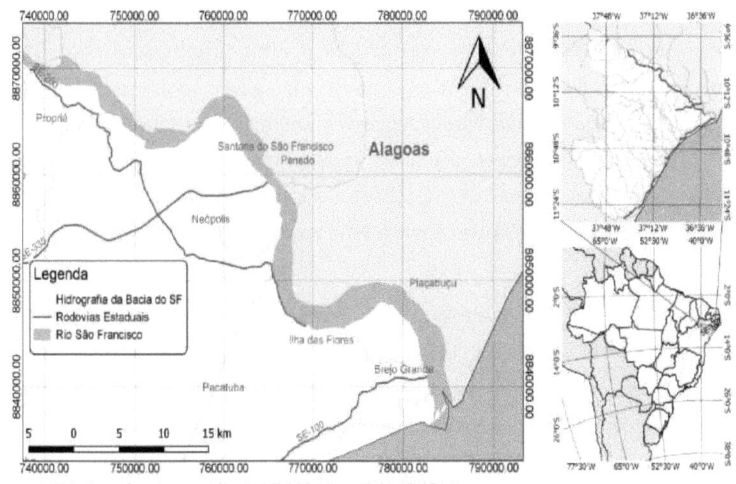

Source: SEMARH/SE

3.2.2 Hydrodynamic model

The hydrodynamic modeling was carried out using the SisBahia system (Sistema Base de Hidrodinâmica Ambiental) version 9.1, which satisfactorily represents the phenomenon of interest. In addition, the interface of this system is considerable, user-friendly and has been constantly improved through research. The main input data for SisBaHiA are: flows, wind speed, river bathymetry, harmonic constants, finite element mesh, land and water contour map, roughness, precipitation, among others. Access to this software is free under a term of responsibility acquired by COPPE/UFRJ, which allows it to be used scientifically and for managing water resources.

In the hydrodynamic model, calibration processes are minimized due to the spatial discretization via finite elements, which maximizes the reliability of the data and can be used to simulate various scenarios involving rivers, estuaries, coastal zones, bays, canals, lakes, lagoons and reservoirs (ROSMAN, 2015).

SisBahia has a hydrodynamic model called FIST3D (Filtered in space and time), optimized for natural bodies of water with a free surface where turbulence modeling is based on filtering techniques, similar to those applied in Large Eddy Simulation (LES), considered the state of the art for geophysical flow turbulence (ROSMAN et al. 2001).

SisBaHiA's FIST3D hydrodynamic model is made up of two modules: the first is vertically or horizontally two-dimensional (2DH), through which the free surface elevation and current velocities (2DH) are calculated vertically, while the second module, called 3D, calculates the three-dimensional velocity field through two possible options.

a)   Fully numerical 3D model, coupled to a 2DH module. FIST3D is a complete 3D model for homogeneous fluids.

b)   3D analytical-numerical model to obtain the velocity profiles in the horizontal flow field. This option is more efficient in computational terms, but only considers the advective acceleration in the 2DH module. It therefore gives less accurate results in regions where the advective accelerations vary significantly along the depth. In this option, the velocity profiles are computed using a solution that is a function of the vertically averaged 2DH velocities, the elevation of the free surface, the equivalent bottom roughness of the 2DH module, and the wind speed acting on the free surface of the water.

The mathematical formulation of the hydrodynamic model comprises the Navier-Stokes equations, which are fundamental for representing any body of water. The simulation results can be represented in 3D or 2DH depending on the input data. 2D models present predominantly two-dimensional flow and require considerable numbers of parameters, which need to be well known in order not to generate inaccurate results. In this research, the 2DH module was used to meet the requested objective. A summary of the 3D governing equations is shown in Figures 28 and 29.

Spatial discretization was carried out using 601 fourth-order quadrangular elements. The vertical discretization of the water column was done using finite differences with sigma transformation. The time step used in the hydrodynamic simulations was 30 seconds, with a maximum Courant equal to 3.0.

Figure 28: Equation of the quantity of motion, with hydrostatic approximation, in the ×· direction

$$\underbrace{\frac{\partial u}{\partial t}}_{(1)} + \underbrace{u\frac{\partial u}{\partial x} + v\frac{\partial u}{\partial y} + w\frac{\partial u}{\partial z}}_{(2)} = \underbrace{-g\frac{\partial \zeta}{\partial x}}_{(3)} - \underbrace{\frac{1}{\rho_0}g\int_z^\zeta \frac{\partial \rho}{\partial x}dz}_{(4)} + \underbrace{\frac{1}{\rho_o}\left(\frac{\partial \tau_{xx}}{\partial x} + \frac{\partial \tau_{xy}}{\partial y} + \frac{\partial \tau_{xz}}{\partial z}\right)}_{(5)} + \underbrace{2\Phi\,\text{sen}\,\theta v}_{(6)}$$

Source: (SILVA, 2014).

Figure 29: Equation of the quantity of movement, with hydrostatic approximation, in the y direction:

$$\underbrace{\frac{\partial v}{\partial t}}_{(1)} + \underbrace{u\frac{\partial v}{\partial x} + v\frac{\partial v}{\partial y} + w\frac{\partial v}{\partial z}}_{(2)} = \underbrace{-g\frac{\partial \zeta}{\partial y}}_{(3)} - \underbrace{\frac{1}{\rho_0}g\int_z^\zeta \frac{\partial \rho}{\partial y}dz}_{(4)} + \underbrace{\frac{1}{\rho_o}\left(\frac{\partial \tau_{yx}}{\partial x} + \frac{\partial \tau_{yy}}{\partial y} + \frac{\partial \tau_{yz}}{\partial z}\right)}_{(5)} - \underbrace{2\Phi\,\text{sen}\,\theta u}_{(6)}$$

Source: (SILVA, 2014).

The details and organization of the formula are based on

(ROSMAN, 2016, p. 45) and (SILVA, 2014, p. 8).

(1)   This is the local acceleration of the flow. In permanent flows, this term is equal to zero.

(2)   It represents the advective acceleration of the flow. In uniform flows, these terms are equal to zero.

(3)   It represents the variation in hydrostatic pressure in the x direction (pressure gradient) due to the slope of the free surface in the x direction. As indicated by the negative sign, this term forces flows from places where the water level is higher to places where the water level is lower.

(4)   It represents the variation in hydrostatic pressure in the x,y direction (pressure gradient) due to differences in density. As indicated by the negative sign, this term forces flow from places where the water is denser to places where the water is less dense.

(5)   It represents the resultant of the dynamic turbulent tensions in the flow (diffusive flow balance). Among other things, it is through these terms that the flow feels the friction of the bottom and the action of the wind on the free surface, generating the velocity profiles.

(6)   It represents the Coriolis acceleration due to the reference frame moving with the Earth's rotation. This term is negligible near the equator.

3.2.3 Initial data for modeling

To draw up the geometric outline of the Sâo Francisco River mouth area, we used the nautical chart of the Sâo Francisco do Norte River mouth bar, drawn up by the Hydrography and Navigation Department (DHN) and coordinates defined using satellite images from the Google Earth program. This stage is considered important in order to structure the spaces and the representation of the simulated environmental phenomena.

The nautical chart was inserted into the Surfer program, version 12, to create the land and sea contours and then imported into the model. The land contour represented the dry part limiting the main banks of the Sâo Francisco, Paraùna and Potengy rivers, as well as the Criminosa, Fitinha and Negra islands, and the Parapuca channel. For the representation of the land and sea contours, the blue color indicates the sea, an open area, and the brown color the land contour Figure 29.

°The information regarding the bathymetry of the study area was taken from the DHN nautical charts (No. 1002 and 22300), bathymetric surveys carried out by the GeoRioMar research department of the Federal University of Sergipe and data provided by the captains of the native boats in the region, data that complemented areas where the nautical charts did not provide information.

The equivalent bottom roughness value ε^5 adopted was 0.020m, with a predominance of fine and medium sands, based on the analysis of the granulometry of the bottom sediments carried out at ITPS (Technological Research Institute of the State of Sergipe), following the values suggested for the effective amplitude of the bottom roughness without wave effects between 0.0070m and

---

5 ε Means amplitude

0.0300m (Abbot and Basco[6] , 1989 apud Rosman, 2015). The bathymetric map shows the depth profile distributed throughout the assessed channel, as shown in Figure 30.

Figure 30: Bathymetry used in the modeling domain.

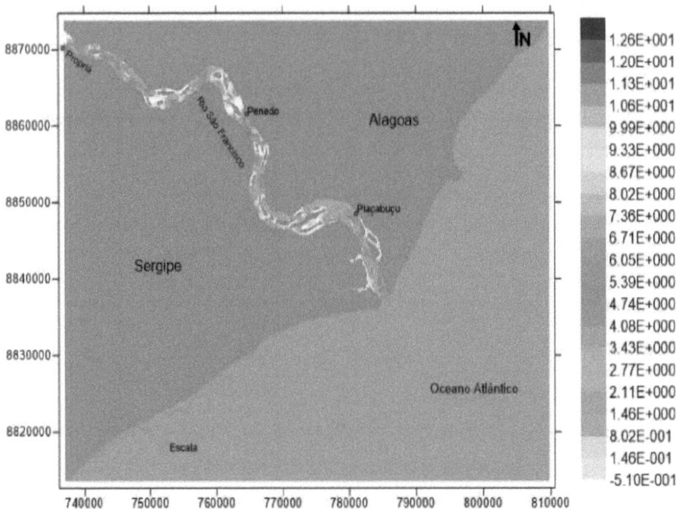

The spatial discretization of the modelling domain was carried out using a finite element mesh, representing the main configurations of the stretch from the river mouth to the town of Propria/SE. The discretization mesh has 1,854 nodes in the horizontal plane and 1,053 vertical levels, totaling 2,907 calculation points, 128.9km^2 of area and an average depth of 4.36m.

3.2.4 Model forces

The harmonic constants used to represent the tidal force were based on the Cabeço tide gauge station (Station 127), located on the Foz do Rio Sao Francisco bar in front of the lighthouse, controlled by the Sea Studies Foundation (FEMAR, 2016). The harmonic constant data is described in the table in the ANNEX.

In order to verify the flow variations and their behavior in the course of the waterbed in the region of the mouth of the São Francisco River, a 39-year time series of river discharges (Q_f) of monthly averages was analyzed. The average flow released by the Xingó hydroelectric power station in the municipality of Propria for the simulated year was considered to be constant at around 952m s$^{3-1}$, from 2015 to 2016, data obtained from the National Water Agency (ANA), Figure 31.

Figure 31: Graph of the average monthly flows in the lower São Francisco River from 1979 to 2016 recorded by the fluviometric station in the municipality of Propriâ/SE.

<hr>

6 Abbot, M.B; Basco, D.R. Computational Fluid Dynamics, an Introduction for Engineers, Longan Group, UK Limited, 1989.

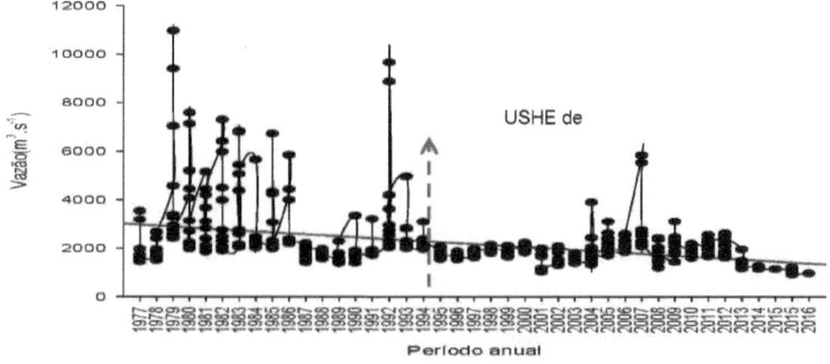

Fonte: Agencia Nacional de Águas (ANA)

3.2.5 Scenario simulations

To monitor the simulated scenarios, six points were defined along the course of the river. Point 1 refers to Ilha Criminosa, approximately 4km from the mouth, Point 2 Piaçabuçu/AL is 10km away, Point 3 is located in the municipality of Ilha das Flores/SE 16km away, Point 4 Penedo/AL 30km away, Point 5 Saùde/SE 37km away, Point 6 Propriâ/SE 57km away from the mouth, as shown in Figure 10. It is worth noting that the points were directed by the location of the river's main channel, resulting in the registration of points in the states of Sergipe and Alagoas, as shown in Figure 32.

Figure 32: Map showing the flow points assessed along the mouth of the River Sâo Francisco.

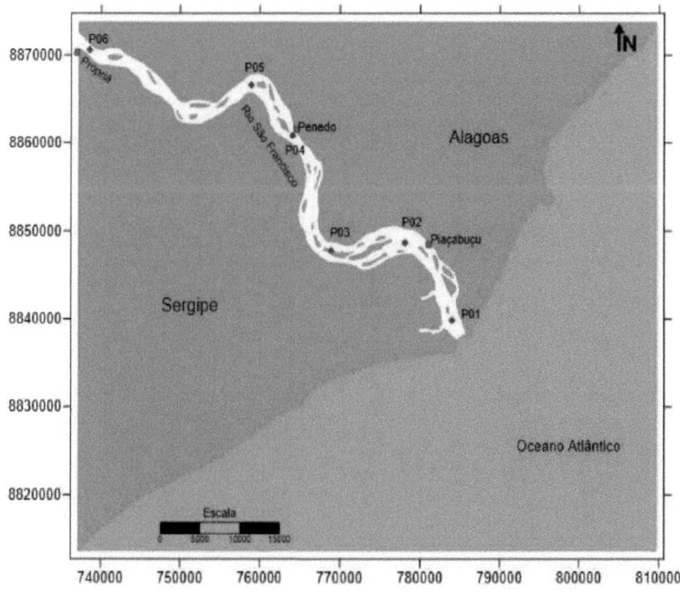

3.3 Results and discussion

3.3.1 Calibration of the hydrodynamic model

To calibrate the hydrodynamic model, the simulations were adjusted based on the parameters of the 6ª field campaign carried out on September 26 and 27, 2015, at the end of the quadrature tide and the beginning of the sizigia tide, when the vertically oscillating waters are at their highest, closing a 12-hour cycle. This cycle well characterized the existing tidal regime at the mouth of the River San Francisco, which is considered to be semidiurnal with an interval of 12.4 hours, a classification detailed by Miranda, Castro and Kjerfve, (2002), through a calculation presented by shape number (N_f) in which the result obtained was 0.16, thus classifying the type of tide in the study area.

Validating the model by checking the tide level variations and then the current speeds is part of the methodology suggested by Rosman et al (2013) in the modeling process. The first checks showed that the bathymetry, velocity and flow rates simulated corresponded to the data measured in the field. The velocity data was taken from the monitoring report of the Companhia Hidroelétrica do Sao Francisco (CHESF, 2009), which carried out monitoring at a point located 4 km from the mouth, with UTM coordinates 24L 784605 and 8841266.

Pearson's correlation was used to analyze the results at the sampling point. França (2013) points out that, after calibrating the model, the errors relating to water levels should be less than 5%, and velocity and flow less than 20%. Based on these values, the correlation was 97.98% and 94.28%, indicating a strong correlation with the simulated data (Figure 33 and 34), for the values recorded in the field and simulated see Table 15 APPENDIX. A total of 8 calibrations were carried out, the last of which was considered to behave adequately and was the best calibration percentage.

Figura 33:   Analysis of the correlation between measured and simulated water current velocity on 17/12/2009 from 15:00 to 22:00.

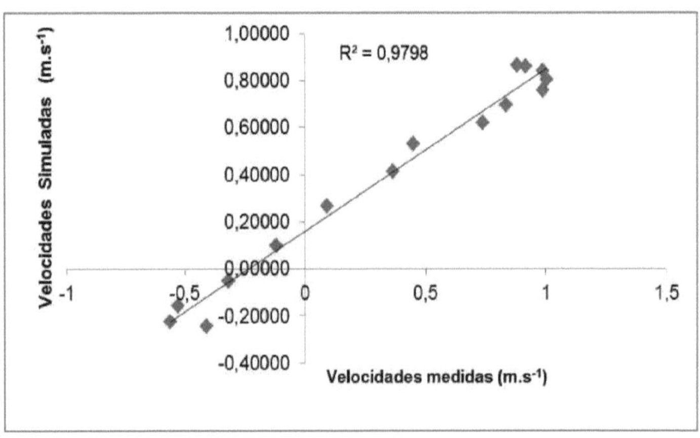

Figura 34:   Analysis of the correlation between measured and simulated water current velocity on

18/12/2009 from 01:00 to 10:30.

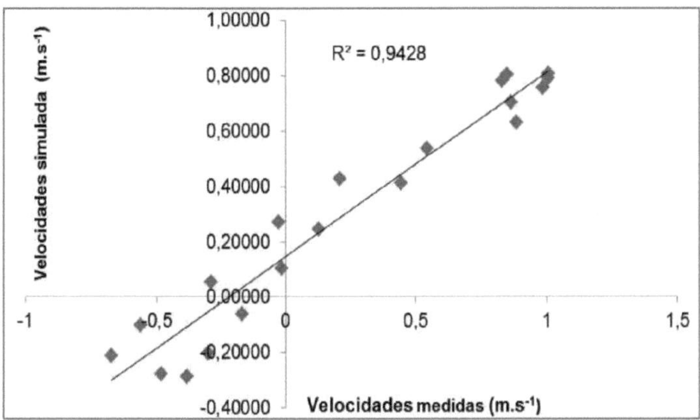

It is understood that the SisBahia hydrodynamic model already contains self-calibration mechanisms, requiring only the input data, providing consistency between the actual and predicted values at 90% water levels and around 70% flow and wind direction. The simulation took place over a period of 13 days, closing a sizzling tide cycle Figure 35.

Figure35: Simulated tides for the period from 16/09/2015 to 28/09/2015.

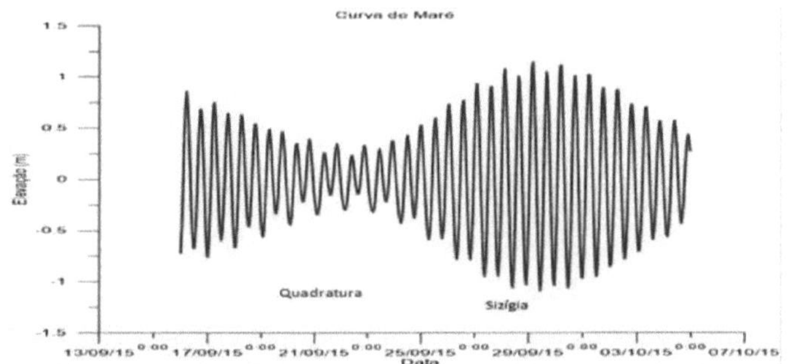

Source: Cabeço/SE Marine Station

3.3.2 Hydrodynamic scenario at the mouth of the São Francisco River

An analysis of the dynamics of the river flow and the expansion of tidal movements in Figures 36F and 36B shows that only at the Proprià and Penedo points are the river discharge values within the range recorded by the fluviometric station, but the other points show values above those recorded, indicating the entry of the marine current into these regions, with the highest incidences at the Ilha Crimosa point.

According to Miranda, Castro and Kjerfve, (2002), the behavior of the meeting of river water with that of the sea carried by the tide extending upstream is an event characteristic of estuarine environments, however, the data reveal the expansion of the influence of the tide to the municipality of Ilha das Flores due to the recording of the fluvial force above $1500m^3 .s^{-1}$.

It can be seen that at Point 1 (Crimean Island), the greatest variations occurred, reaching $-6000 m^3 .s^{-1}$ at flood tide and $-4000 m^3 .s^{-1}$ at flood tide. At another point on the ebb tide, the flows ranged from $8000 m^3 .s^{-1}$ on the spring tide to $5595 m .s^{3-1}$ on the ebb tide. Point 1 (Criminal Island) is the most representative of the tidal ranges among the other points evaluated, reinforcing the dominance of the tidal flow over the river flow. Results contained in the monitoring report carried out by CHESF in 2009 in the region of the mouth of the River Sâo Francisco showed average flows of around $2,097 m^3 .s^{^1}$, during the period of the quadrature and sizigia tides. These levels prove that the river flow reacts to the expansion of the tidal propagation in the estuary.

However, today, with regularized flows of around $900 m^3 .s^{-1}$, the scenario is reversed. According to CHESF (2015), the practice of $900 m^3 .s^{-1}$ was necessary due to the lowering of the Sobradinho/BA reservoirs. According to Medeiros et al (2007), the post-dam effect made it possible to regularize and decrease the magnitude of the flows along the Sâo Francisco River, drastically reducing its flood pulses. There is also tidal influence at Point 2 (Piaçabuçu/AL) and Point 3 (Flores Island/SE), but to a lesser extent than at Point 1.

It can be seen that in P6 (Propria/SE) the variations in the quadrature and sizigia tides range from 930 to $970 m^3 .s^{^1}$, reaffirming the phenomenon of regularization of flows maintained by the Xingó HPP. Figure 37 shows the speed of currents in the water column, where the average value ranged from $0.04 m.s^{-1}$ in the municipality of Penedo/AL to $0.14 m.s^{-1}$ in the region of Propria/SE. The maximum values for the ebb period were $0.17 m.s^{-1}$ and for the flood period $0.19 m.s^{-1}$.

Throughout the simulations for the pre-tidal hydrodynamic circulation, well-defined vortices formed by the speed of the circulation of the marine current were seen above the town of Piaçabuçu/AL, bordering below the town of Penedo/AL (Figures 36 and 37). At low tide, the vortices indicate a linear flow towards the sea, a behavior which indicates that the low tidal oscillations do not interfere with the fluvial force in the main stretch of the Foz channel, but vortices in the direction of the continent can be seen entering the Parapuca channel. Roversi, Rosman and Harari (2016), in an analysis of hydrodynamics in the Santos estuary, found that the most intense velocities occurred at high tide, indicating the predominance of the effect of the tide on the renewal and exchange of water in the system.

In the flood simulations, it can be seen that in the middle of the municipality of Penedo/AL there is a limitation of the vortices related to the action of the marine current, this scenario indicates a stability between the river pulse and the strength of the marine current (Figure 36). However, with the ebb simulation, the vortices from the action of the marine current are not evident (Figure 37). According to Miranda, Castro and Kjerfve (2002), the mixing zone is the region where freshwater and seawater mix. These factors prove that the hydrodynamics of an estuary result from the interaction between the morphology of the estuarine basin, the fluvial input and the tidal regime (D'AQUINO et al 2011b).

Figure 36: Simulation of the hydrodynamics of flood velocity~(A). (B) Velocity above the municipality of Saùde and below the municipality of Propria/SE.

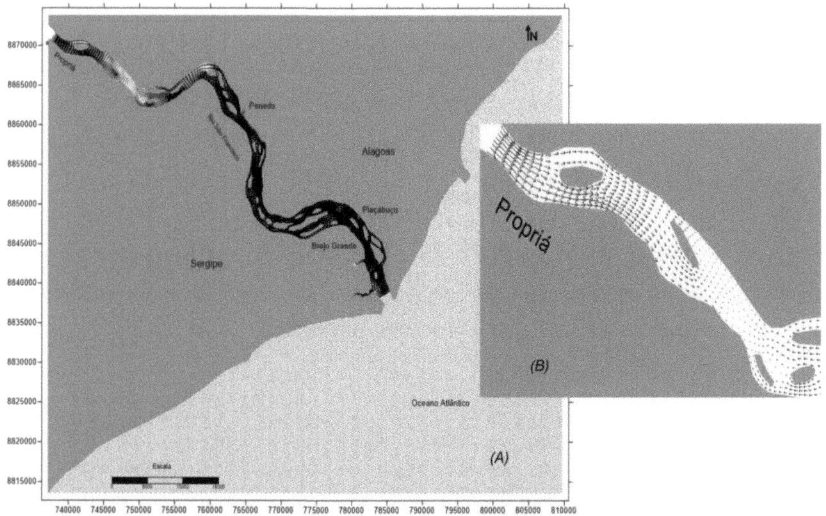

Figure 37: Simulation of the hydrodynamics of the ebb velocity (A). (B) Foz to Piaçabuçu velocity.

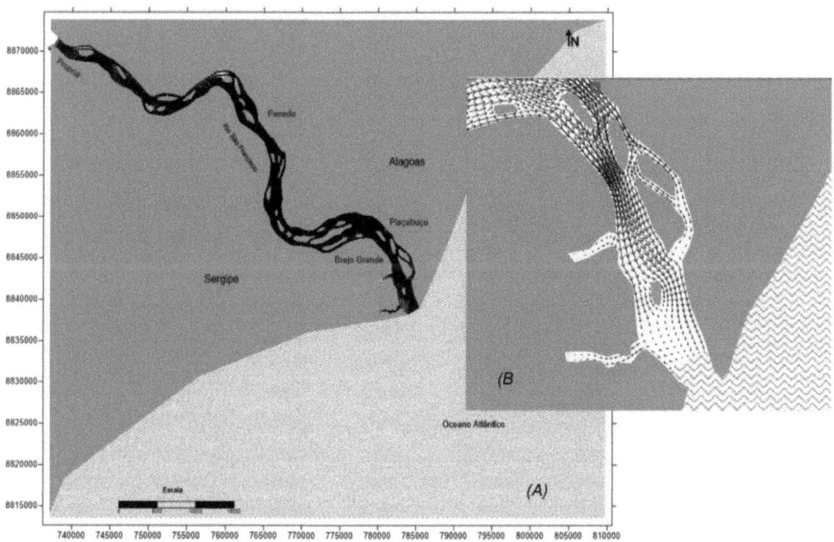

4 CONCLUSION

In conclusion, the use of the environmental hydrodynamic model made it possible to understand the existing hydrodynamics at the mouth of the River San Francisco. The simulations showed the dynamics of the flow and the velocity profile during the river's ebb and flood periods.

77

The hydrodynamic scenario showed the influence of the marine current on the river flow in stretches above Piaçabuçu/AL, an occurrence that could lead to changes in environmental dynamics and alterations in water quality characteristics, to the detriment of the reduction in the useful volume of the existing reservoirs downstream of the São Francisco River.

References

ARAGÃO, T. N. K. C.; SEVERIANO, J. S.; MOURA, A. N. Phytoplankton composition of the Itaparica and Xingó reservoirs, Sao Francisco River, Brazil **.Braz. J. Biol.**, 2015, vol. 75, no. 3, p. 616-627

BORTONE, F. A. S.; LUDWIG, M. P.; XAVIER, K. D. Contradictions of modernity in the process of de/re/ territorialization of place: The case of those affected by the construction of the Candonga Hydroelectric Plant. **Revista ELO - Dialogos em Extensao** Volume 05, number 02 - October 2016.

BRAZIL. National Department of Mineral Production. Ordinance n. 416, 03/09/2012. Available at: <http://www.dnpm.gov.br/acesso-a-

informacao/legislação/portarias-do-diretor-geral-do-dnpm/portarias-do-diretor-   geral/portaria-no-416-em-03-09-2012-do-diretor-geral-do-dnpm>. Accessed September 05, 2017.

CASTRO, P.; HUBER, M. E. Marine biology. 8" ed. AMGH. 2012. 436p.

CAVALCANTE, G. MIRANDA, L. B.; MEDEIROS, P. R. P. Circulation and salt balance in the Sao Francisco river Estuary (NE/Brazil). Revista Brasileira de Recursos Hidricos Brazilian Journal of Water Resources. On-line version ISSN 2318-0331 RBRH, PortoAlegre, v. 22, e31, 2017

CBHSF - São Francisco River Basin Committee. 2015. Ten-Year Water Resources Plan for the São Francisco River Basin - PBHSF (2016-2025) In: Project for integrated management of activities carried out on land in the São Francisco Basin - Synthesis of the Executive Summary of the PBHSF with Consideration of CBHSF Deliberations. Brasilia: ANA, 274 p.

CBHSF - São Francisco River Basin Committee. 2015. Contingency Plan for Flexibility of the Minimum Restriction Flow to 800 m^3 /s in the São Francisco River Basin. Part I: Multiple Uses. 2015.   15   p.   Available at:<http://arquivos.ana.gov.br/saladesituacao/ReducaoTemporaria/Outros/201 5/PlanoContingenciaParaFlexibilizacaoVazaoMinimaRestricaoPara_800m3_Ba ciaRioSaoFrancisco_Parte_1_MeioAmbiente.pdf> Accessed on: 13.09.2017

CBHSF - São Francisco River Basin Committee. Water Resources Plan for the São Francisco River Basin. Monitoring and Evaluation of the Introduction of the Saline Wedge into the São Francisco Estuary. 2009. Available at:<

http://licenciamento.ibama.gov.br/Recursos%20Hidricos/Integracao%20Sao%2 0Francisco/Relatorios%20execucao%20PBA/REL%20SEMESTRAL%207/Prog rama%2036/Anexo%204.36.1 %20Cunha%20Salina%20Jan-Jul09.pdf> Accessed on September 20, 2016.

D'AQUINO, C. A.; ANDRADE NETO, J. S.; BARRETO, G. A. M.; SCHETTINI, C. A. F.; Oceanographic characterization and transport of suspended sediments in the estuary of the Mampituba River, SC. **Revista Brasileira de Geofisica**. Vol 29(2), p.217-230. 2011a.

D'AQUINO, C.A; FRANKLIN DA SILVA, L.; COUCEIRO, M.A.A; PEREIRA, M.D. Salt Transport and Hydrodynamics of the Tubarao River Estuary - SC, Brazil RBRH. **Revista Brasileira de Recursos Hidricos**. Volume 16 n.3 - July/September 113-125p, 2011b.

DERROSSO, G.; ICHIKAWA, E. Y. O papel da Crabi no assentamento dos ribeirinhos atingidos pela construgao da hidrelétrica de Salto Caxias no estado do Parana. **Rev. Adm. Pùblica** - Rio de Janeiro 47(1):133-155, Jan./Feb. 2013.

DYER, K. R., Sediment transport processes in Estuaries: in: Geomorphology and Segimentology of Estuaries. G.M.E. Perillo (Ed) development in Sedimentology, 53 ElsevierScience. 1995 p.423-449

FEMAR. Sea Studies Foundation. Estagao Maregrafica Brasileira. 20016. Available at https://www.fundacaofemar.org.br/biblioteca/emb/Tabelas/126.html. Accessed on January 13, 2016.

FONTES, L. C. SILVEIRA. FROM SOURCE TO BASIN: CONTINENT-OCEAN INTERACTION IN THE SAO FRANCISCO RIVER SEDIMENTARY SYSTEM, BRAZIL. Thesis (doctorate) - Universidade Estadual Paulista, Instituto de Biociencias de Rio Claro. 315p, 2016.

FONTES, L. C. SILVEIRA. From source to basin: continent-ocean interaction in the são francisco river sedimentary system, brazil. Thesis (PhD) - Universidade Estadual Paulista, Instituto de Biociencias de Rio Claro. 2016, 315p

FRANÇA, J. M. B. Evaluation of the degradation of the Acarape do Meio-Ce reservoir using mathematical modeling and geotechnology. Dissertation. Water Resources Management, Department of Hydraulic and Environmental Engineering, Federal University of Ceara. 106p. 2013.

HANSEN D. V.; RATTRAY M. 1966. New dimensions on estuarine classification. LimnologyandOceanography,11: 319-326.

IPEA. Institute for Applied Economic Research. Irrigation Project in the Lower Sao Francisco Valley. Brasilia: Ipea, 1992, 44p. Available at http∜repositorio. ipea.gov.br/bitstream/11058/2517/1/TD%20268.pdf Accessed on March 1, 2016.

LERCZAK, J. A.; GEYER, W. R.; CHANT, R. J. Mechanisms driving the time dependent salt flux in a partially stratified estuary. **Journal of Physical Oceanography**, v. 36, n. 12, p. 2296-2311, 2006. http:ZZdX.doi.orgZ10.1175ZJPO2959.1.

LEWIS, R. E.; LEWIS, J. O. The principal factors contributing to the flux of salt in a Narrow, partially Stratified Estuary. **Estuarine and Coastal Marine Science**, v. 16, n. 6, p. 599-626, 1983. http√/dx.doi.org/10.1016/0272- 7714(83)90074-4.

LOITZENBAUER, E. MENDES, C. A. B. Salinity dynamics as a tool for the integrated management of water resources in the coastal zone: an application to the Brazilian reality **Revista da Gestao Costeira Integrada** 11(2):233-245 (2011).

MARTINS, D. M. F.; CHAGAS, R. M.; MELO NETO, J. O.; MELLO JUNIOR, A. V. Impacts of the construction of the Sobradinho hydroelectric plant on the flow regime in the Lower São Francisco River. **R. Bras. Eng. Agric. Ambiental**, v.15, n.9, p.1054-1061, 2011.

MEDEIROS P.R.P., Knoppers B.A., dos Santos Jr R.C., Souza W.F.L. Fluvial transport and dispersion of suspended particulate matter in the coastal zone of the São Francisco River (SE/AL). **Geoch Bras**, 209-228. 2007.

MEDEIROS P.R.P., Knoppers B.A., Souze W.F.L., Oliveira E.N. 2011. Transport of suspended material in the lower Sao Francisco River (SE/AL) under different hydrological conditions. **Braz J Aquat Sci Technol**, 15:42-53

MEDEIROS, P. R. P. Fluvial transport, transformation and dispersion of suspended matter and nutrients in the estuary of the São Francisco River, after the construction of the Xingó Hydroelectric Power Plant (AL/SE). Doctoral thesis. Department of

Geoquimica, Universidade Federal Fluminense, 2003.184 p.

MIRANDA, L.B.; CASTRO, B.M.; KJERFVE, B. **Principios de oceanografia fisica de estuãrios.** EDUSP - Editora da Universidade de Sao Paulo. Sao Paulo, Brazil. 424 p. 2002.

MONTEIRO, S. M; EL-ROBRINI, M; CASTRO ALVES, I. C. Seasonal dynamics of nutrients in an Amazon estuary. Mercator, Fortaleza, v. 14, n. 1, p. 151162, jan./abr. 2015.

PEREIRA, Marçal D.; SIEGLE, Eduardo; MIRANDA, Luiz B. de and SCHETTINI, Carlos A.F..Hydrodynamics and seasonal suspended particulate matter transport in a tidal estuary: Caravelas Estuary (BA). **Rev. Bras. Geof.** [online]. vol.28, n.3, 427-444p. 2010.

PEREIRA, Marçal D.; SIEGLE, Eduardo; MIRANDA, Luiz B. de and SCHETTINI, Carlos A.F..Hydrodynamics and seasonal suspended particulate matter transport in a tidal estuary: Caravelas Estuary (BA). *Rev. Bras. Geof.* [online]. 2010, vol.28, n.3, pp.427-444.

RIGO, D. Analysis of the flow in estuarine regions with mangroves - measurements and modeling in the bay of Vitória/ES. Thesis. Federal University of Rio de Janeiro/COPPE. 2004. 170p.

ROSMAN, P.C.C. A computational system for environmental hydrodynamics. In: Silva, R.C.V. (Ed.), Numerical Methods in Water Resources - Volume 5. ABRH. p. 1-161.2001.

ROSMAN, P. C. C., (Ed.). SisBaHiA® technical reference. COPPE - Federal University of Rio de

Janeiro, March 2015, Available at: <http://www.sisbahia.coppe.ufrj.br/SisBAHIA_RefTec_V95. pdf> Accessed on: Jun. 2016.

ROSMAN, P. C. C., CUNHA, C. L. N., CABRAL, M. M. et al. SisBaHiA Technical Reference. COPPE /UFRJ. Ocean Engineering Program, Rio de Janeiro, RJ. Version 9.2,   2013.   249   p. Available at:

http://www.sisbahia.coppe.ufrj.br/SisBAHIA_RefTec_V92.pdf. Accessed on: September 20, 2015.

ROVERSI, F.; ROSMAN, P. C.; HARARI, J. Analysis of the Trajectories· of Continental Waters Flowing into the Santos Estuarine System. **RBRH** vol. 21 no.1 Porto Alegre jan./mar. p. 242 - 250.2016.

SANTOS, E. S. Hydrodynamic modeling and water quality in a pororoca region at the mouth of the Araguari River/AP. Dissertation. Federal University of Amapa Foundation. 2012. 113p.

SEGUNDO, G. H. C. Sedimentological hydrodynamic characterization of the estuary and delta of the São Francisco River. Dissertation Meteorology Department/CCEN/UFAL. 2001. 102p

SHI, J. Z.; LU, L. F. A short note on the dispersion, mixing, stratification and circulation Withinthe plume of the partially-mixed Changjiang River estuary, China. **Journal of Hydro-environment Research.** p 111-112, 2011.

CHAPTER 4: SALINITY IN THE FLOOD OF THE SĂŁO FRANCISCO RIVER USING COMPUTER MODELING

SUMMARY

Salinity is a physical-chemical property that represents the concentration of dissolved salts in a body of water. The study was carried out in the Foz do Rio Sao Francisco area between the municipalities of Propria/SE and Piaçabuçu/AL, regions belonging to the Rio Sao Francisco hydrographic basin. The hydrodynamic model was developed through computer simulations carried out using the SisBAHIA system (Sistema Base de Hidrodinâmica Ambiental) model 9.1. The salinity simulation was based on the sizzling tide, while the validation was based on field data measured with a multi-parameter probe in September 2015. After validating the salinity, we began simulating scenarios to analyze the behavior of salinity dynamics along the mouth of the San Francisco River. The scenarios obtained indicated that salt transport is directed by the force of the seas, which is accentuated at the entrance to the mouth of the River San Francisco, reaching values of 32‰ at high tide. Another aspect presented in the simulation is the distribution of salinity between the tributaries on the left bank located in Brejo Grande/SE, such as the Paraùna river and the Parapuca channel.

Keywords: Salinity; Flow; Sea regime.

4.1 Introduction

Salinity is a physical-chemical property that represents the concentration of dissolved salts in a body of water. In aquatic environments close to the coastal zone, the concentration of salts present in the water can vary according to marine oscillations.

Salinity in estuaries varies greatly due to hydrodynamic processes, river inputs and the exchange of estuarine water with the atmosphere (MIRANDA, CASTRO and KJERFVE, 2002).

The main physical phenomenon in an estuarine region is the mixing of salt water and fresh water. The dynamics of this process generate a salinity gradient that runs from the west towards the ocean (LOITZENBAUE and MENDES 2011). Along the Brazilian coast there are several rivers that discharge their waters into the oceans, the Sâo Francisco River is one of them, its mouth divides the state of Sergipe and Alagoas and is considered a unique estuarine system with natural beauty being driven by tidal regimes.

Faced with the advance of the sea and decreasing flows, the mouth of the River San Francisco is facing the salinization of its surface waters caused by the advance of the saline wedge towards the upstream end of the river. The decrease in the tributary flow has increased the salinization of the waters at the mouth of the River San Francisco, causing impacts on public supply, agriculture and fishing.

One of the reasons for the decrease in tributary flow is the construction of dams along the course of the river, which limit the flow of fresh water into the ocean, increasing the concentration of salts in the estuary.

According to Medeiros et al (2007), one of the most notable modifications of the construction of dams on rivers is the regularization of the flow, with the aim of supplying the water needed to generate electricity, causing a great reduction in the natural flow, causing an imbalance of energy between the river and the sea.

The aim of this study is to verify the dynamics of salinity in the area of the mouth of the River San Francisco using the SisBAHIA hydrodynamic model.

4.2 Methodology .

4.2.1 Study area

The São Francisco river basin has a drainage area of $639,219km^2$ and a main riverbed 2,700km long with an average flow of $2,850m^3/s$. The source of the São Francisco River is located in the Serra da Canastra in Minas Gerais, running through the states of Bahia, Fernambuco, Alagoas, Sergipe, Goias and the Federal District, linking Brazil from the Southeast to the Northeast, representing 7.5% of the country's territory (AGUIAR NETTO et al 2011). According to the National Water Agency (2005), this basin is divided into four physiographic regions: Alto, Mèdio, Submédio and Baixo Sao Francisco which, for planning purposes, these areas were subdivided into thirty-four small basins, and 12,821 micro-basins with the aim of delineating by stretches the main rivers of the region.

The study was carried out in the area of the mouth of the River San Francisco between the municipalities of Propria/SE and Fiaçabuçu/AL (Figure 38), regions belonging to the Lower San Francisco. The region's climate is divided into tropical semi-arid and tropical semi-humid, with evapotranspiration ranging from 600 to 700mm in the wet season and 750mm to 800mm in the dry season. Annual rainfall may decrease from the coast, there are hydromorphic soils, including organic, gley and alluvial soils, which in their natural state are subject to periodic flooding, have limited fertility and are more suitable for rice cultivation (IFEA, 2002).

The coastal strip dives under the ocean and advances as the substrate of the continental edge containing stratigraphic units deposited from the Paleozoic to the present day, and the rocks of the sedimentary basin constitute the geological substrate of the final stretch of the Sao Francisco river-sea system (FONTES, 2015).

Figure 38: Map of the São Francisco river basin.

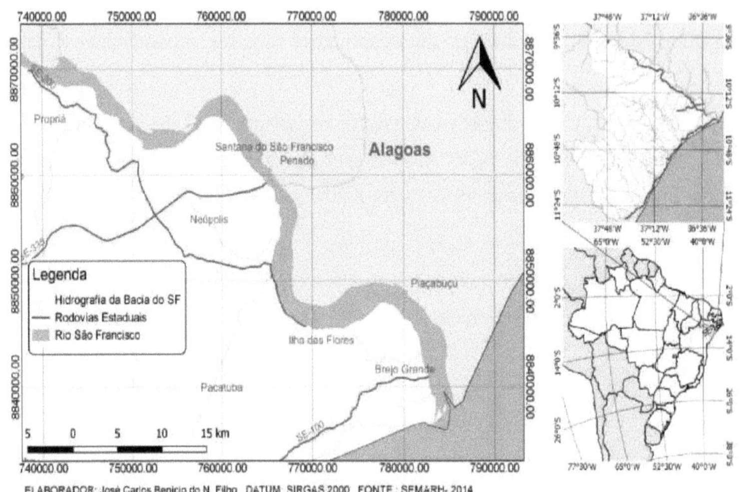

Source: SEMARH/SE

4.2.2 Hydrodynamic model and 2DH

Hydrodynamic modeling was carried out using the SisBAHIA system (Sistema Base de Hidrodinâmica Ambiental) version 9.1, which satisfactorily represents the phenomenon of interest. In addition, the system's interface is considered user-friendly and has been constantly improved through research. The main input data for SisBAHIA are: flows, wind speed, river bathymetry, harmonic constants, finite element mesh, land and water contour map, roughness, precipitation, among others. Access to this software is free under a term of responsibility acquired by COPPE/UFRJ, allowing it to be used scientifically and for managing water resources.

In the hydrodynamic model, the calibration processes are minimized due to the spatial discretization via finite elements, which maximizes the reliability of the data and can be used to simulate various scenarios involving rivers, estuaries, coastal zones, bays, channels, lakes, lagoons and reservoirs (ROSMAN, 2015).

SisBAHIA has a hydrodynamic model called FIST3D (Filtered in space and time), optimized for natural bodies of water with a free surface where turbulence modeling is based on filtering techniques, similar to those applied in Large Eddy Simulation (LES), considered the state of the art for geophysical flow turbulence (ROSMAN et al. 2001).

SisBAHIA's FIST3D hydrodynamic model consists of two modules: the first is vertically or horizontally two-dimensional (2DH), through which the vertical free surface elevation and current velocities (2DH) are calculated, while the second module, called 3D, calculates the three-dimensional velocity field through two possible options.

a)   Fully numerical 3D model, coupled to a 2DH module. FIST3D is a complete 3D model for

84

homogeneous fluids.

b)   3D analytical-numerical model to obtain the velocity profiles in the horizontal flow field. This option is more efficient in computational terms, but only considers the advective acceleration in the 2DH module. It therefore gives less accurate results in regions where the advective accelerations vary significantly along the depth. In this option, the velocity profiles are computed using a solution that is a function of the vertically averaged 2DH velocities, the elevation of the free surface, the equivalent bottom roughness of the 2DH module, and the wind speed acting on the free surface of the water.

The mathematical formulation of the hydrodynamic model comprises the Navier-Stokes equations, which are fundamental for representing any body of water. The simulation results can be represented in 3D or 2DH depending on the input data. 2D models predominantly show two-dimensional flow and require considerable numbers of parameters, which need to be well known so as not to generate inaccurate results. In this research, the 2DH module was used to meet the requested objective.

Spatial discretization was carried out using 601 fourth-order quadrangular elements. The vertical discretization of the water column was done using finite differences with sigma transformation. The time step used in the hydrodynamic simulations was 30 seconds, with a maximum Courant equal to 3.0.

4.2.3 Initial data for modeling

In order to draw up the geometric outline of the area of the mouth of the River San Francisco, we used the nautical chart of the mouth of the river in a northerly direction, drawn up by the Department of Hydrography and Navigation (DHN) and coordinates defined using satellite images from the Google Earth program. This stage is considered important in order to structure the spaces and the representation of the simulated environmental phenomena.

The nautical chart was inserted into the Surfer program, version 12, to create the land and sea contours and then imported into the model. The land contour represented the dry part bordering the main banks of the São Francisco, Paraùna and Potengy rivers, as well as the Criminosa, Fitinha and Negra islands, and the Parapuca channel. °The information regarding the bathymetry of the study area was taken from the DHN nautical charts (No. 1002 and 22300), bathymetric surveys carried out by the GeoRioMar research department of the Federal University of Sergipe and data provided by the boat captains, natives of the region, which complemented areas that the nautical charts did not provide information on.

The equivalent bottom roughness value ε^{7} adopted was 0.020m, with a predominance of fine and medium sands, based on the analysis of the granulometry of the bottom sediments carried out at

---

7 ε Means amplitude

ITPS (Technological Research Institute of the State of Sergipe), following the values suggested for the effective amplitude of the bottom roughness without wave effects between 0.0070m and 0.0300m (Abbot and Basco[8] , 1989 apud Rosman, 2015). The bathymetric map shows the depth profile distributed throughout the assessed channel, as shown in Figure 39.

Figure 39: Bathymetry used in the modeling domain for the hydrodynamics of the mouth of the Sâo Francisco River.

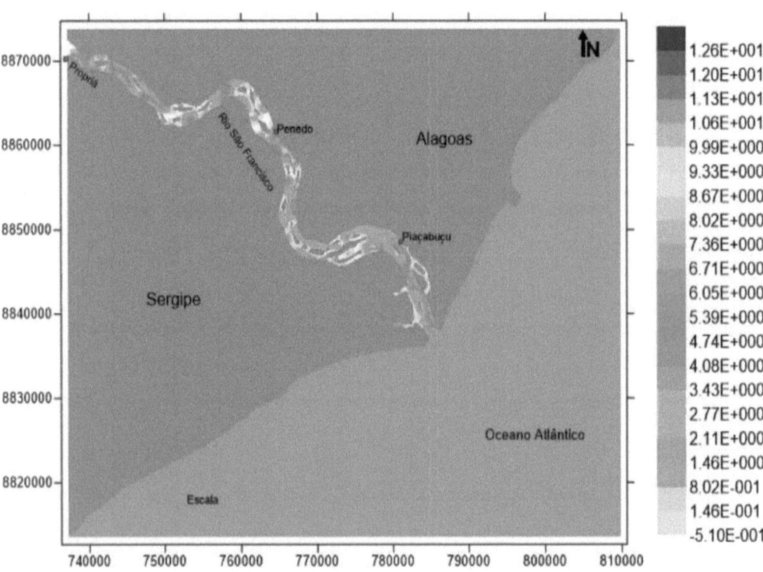

The spatial discretization of the modelling domain was carried out using a finite element mesh, representing the main configurations of the stretch from the mouth to the town of Propria/SE. The discretization mesh has 1,854 nodes in the horizontal plane and 1,053 vertical levels, totaling 2,907 calculation points, 128.9 km^2 of area and an average depth of 4.36m.

Figure 40 also shows the representation of the land and sea contours, with the blue color indicating the sea, an open area, and the orange color the land contour, the white part of the map indicating the modeling domain.

Figure 40: Finite element mesh of the modeling domain for the hydrodynamics of the mouth of the São Francisco River.

---

8 Abbot, M.B; Basco, D.R. Computational Fluid Dynamics, an Introduction for Engineers, Longan Group, UK Limited, 1989.

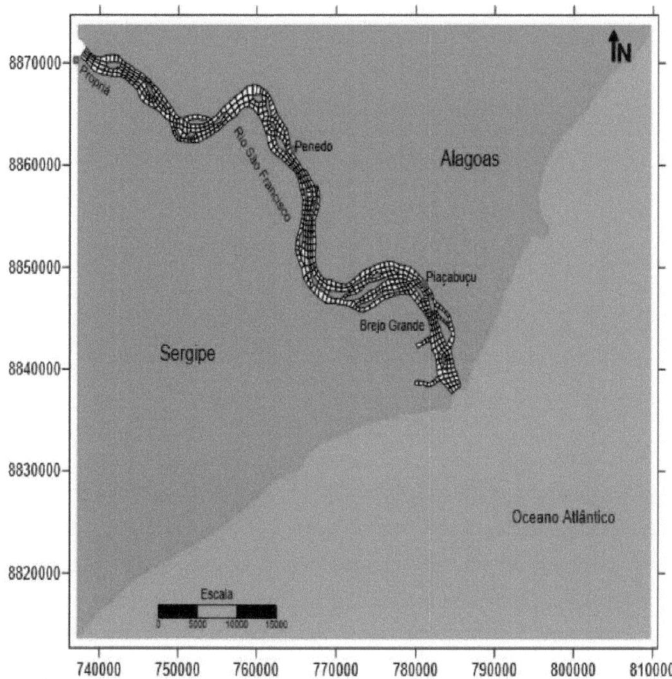

4.2.4 Model forces

The harmonic constants used to represent the tidal force were based on the Cabeço tide gauge station (Station 127), located at the mouth of the São Francisco River opposite the lighthouse, controlled by the Foundation for Marine Studies (FEMAR, 2016). The harmonic constant data is described in the table in the APPENDIX.

In order to verify the flow variations and their behavior in the course of the waterbed in the region of the mouth of the São Francisco River, we analyzed a 37-year time series of the maximum, average and minimum monthly average river discharges (Q_f), data obtained from the National Water Agency (ANA) (Figure 41).

Figure 41: Graph of the annual cycle of average, minimum and maximum flows in the lower São Francisco River from 1979 to 2016, recorded by the fluviometric station in the municipality of Propriâ/SE.

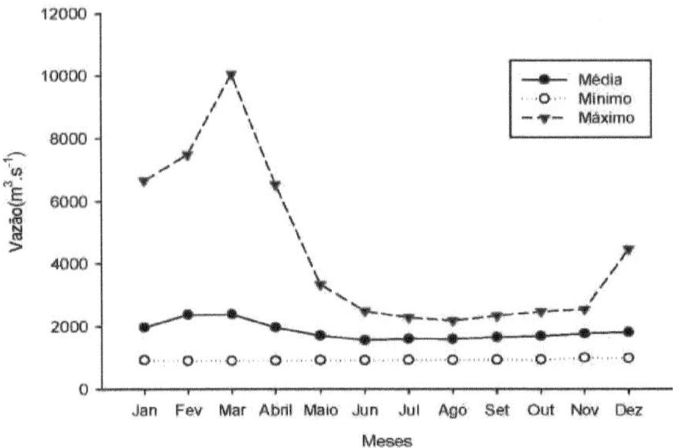

Source: National Water Agency (ANA)

4.2.5 Scenario simulations

To monitor the simulated scenarios, six points were defined along the course of the river. Point 1 refers to Ilha Criminosa, approximately 4km from the mouth, Point 2 Piaçabuçu/AL is 10km away, Point 3 is located in the municipality of Ilha das Flores/SE 16km away, Point 4 Penedo/AL 30km away, Point 5 Saùde/SE 37km away, Point 6 Propria/SE 57km away from the mouth, as shown in Figure 10. It is noteworthy that the points were directed by the location of the river's main channel, resulting in the registration of points in the states of Sergipe and Alagoas, as shown in Figure 42.

Figure 42: Map showing the flow points assessed along the mouth of the River San Francisco.

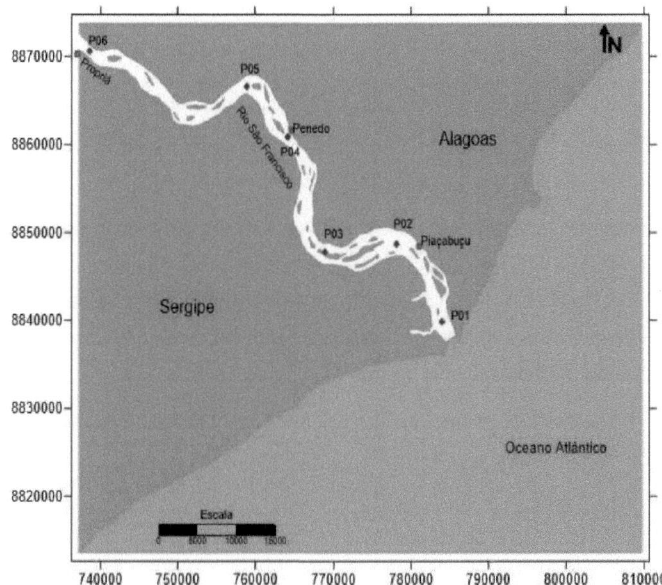

The salinity field data used in this study were obtained from a campaign carried out on September 26 and 27, 2015, at 12-hour intervals, representing the end of the quadrature period and the beginning of the sizigia. The data was collected using a fixed boat 7 km from the mouth of the river and using a HI9828 Hanna multi-parameter probe, recording every 10 seconds at a vertical interval of 1 to 1.5 meters in the water column. Surface salinity profiles along the mouth of the São Francisco River were also evaluated from 2014 to 2015, at an interval of 1km starting at the entrance to the mouth of the river up to a distance of 12km, between the municipalities of Brejo Grande/SE and Piaçabuçu/AL. These data were not used in the modeling because they do not delineate a tidal cycle, but are only representative for understanding the dynamics of salinity displacement in the study area.

For the salt transport simulation, 35‰ was adopted for the open contour area, limiting the sea, and 0.3‰ for the tributaries, representing the river flow. According to Castro and Huber (2012), salinity decreases as it moves upstream and Lewis and Lewis (1983) add that salt transport is correlated with tidal fluctuations. Lerczak (2006) stresses that the amplitude of the oscillation in salt flow is mainly due to advection and advective flows, which are related to the force exerted on the entry and exit of salinity, this scenario being a competition between longitudinal flows.

D'Aquino et al (2011) point out that seawater is denser and flows along the bottom of the watercourse, a phenomenon known as the salt wedge, while freshwater flows in the surface layer towards the sea, with a difference in density. In this way, the values of 35‰ were distributed in all the nodes representing the open border (sea) in the model simulation.

The water column velocity data was taken from the CHESF report (2011), using a current meter with a Savonius propeller, make and model Mini-digi-Kartran, indicating velocities of 0.01 m/s, equipped with a hydrologia brand hydrometric winch with 20 m of cable and a 25 kg current deflector. Wind data was not used in the simulation.

4.3 Results and discussion

4.3.1 Salinity model validation for the Sao Francisco estuary area

To validate salinity, the simulations were adjusted based on data from the point measured in the field, located near the extremities of the municipality of Brejo Grande/SE and Piaçabuçu/AL, with geographic locations of 781214 and 8840694 UTM, which is a reference for salinity validation, where the correlation between the measured and simulated values was 96.05%, with acceptability considered relevant within the statistical foundation, as shown in Figure 43.

Figure 43: Graph of measured and simulated salinity at the reference point for data validation.

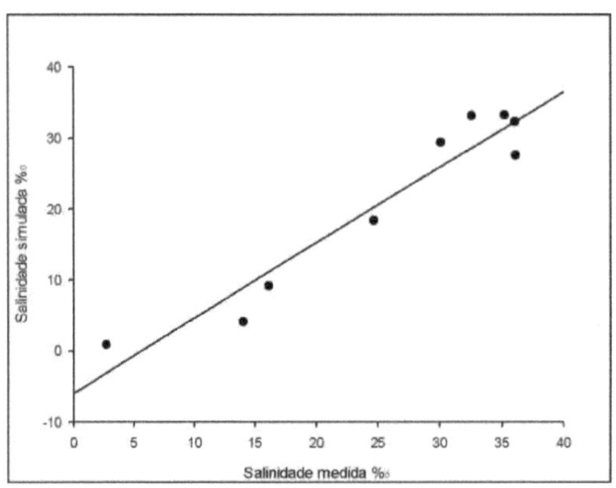

4.3.2 Salinity scenarios at the mouth of the São Francisco River

After validating the salinity, we began simulating scenarios to analyze the behavior of salinity dynamics along the mouth of the São Francisco River. The scenarios obtained began at the mouth of the river up to the municipality of Propria/SE, with the aim of verifying the expansion of the salt wedge.

The salinity profile along the mouth of the River San Francisco varied between 0.05% and 33.56‰ during the spring tide and the water level varied between 0.5m and 2.3m at the points evaluated. The average, maximum, minimum and standard deviation values are shown in Table 15.

Table 15: Measured and simulated salinity values at the Reference Point.

Points	Average	Standard Deviation	Minimum	Maximum
P1 (Salinity ‰)	10,99	10,45	0,69	33,56
P2 (Salinity ‰)	0,27	0,10	0,07	0,72

P3 (Salinity ‰)	0,28	0,04	0,12	0,30
P4 (Salinity ‰)	0,30	0.01	0,25	0,30
P5 (Salinity ‰)	0,30	0,00	0,29	0,30
P6 (Salinity ‰)	0,30	0,00	0,30	0,30

Font

e:Author(2017)

According to Figure 44 (A) at P1 the concentrations reached values of 33.56‰, with variations accompanied by the rhythms of the tidal pulse, canceling out the action of river discharge. According to Miranda, Castro and Kjerfve (2002), in coastal areas dominated by river discharge, the process of entrainment allows salinity to reach the surface layer, in response to tidal and flow variations.

The average salinity for Point 1 is 10.99‰, and it can be seen that the distribution of salinity on 20/09, the quadrature period, ranged from 20‰ to 25‰, increasing to 32.00% at the change of tide. In the initial moments, there is a change from fresh to brackish water, where the salinity indices are <30‰ according to CONAMA resolution 357/2005 and with a polyhaline characteristic with salinity values between 18‰ <30‰ (VELINE, 1958). With this response, the predominant characteristic was brackish water in accordance with CONAMA resolution 357/2005 and with records entering saline characteristics, which can interfere with the aquatic environment and water uses.

Figure 44(B) illustrates the mixing zone at Point 1 with a rise in the water level varying between 0.5m and 2.3m with a longitudinal increase in the frequency of salinity. There is a predominance of saline water in the cross section, contributing to the mixing and homogenization of salinity. At Point 2 in Piaçabuçu/AL, there was an increase in salinity with peaks of 0.72% and a rise in the water level of 0.8m to 1.9m, while at Point 3, in the municipality of Ilha das Flores/SE, the value corresponded to 0.30%, and it should be noted that these values occurred at sizig tide (Figures 45(A) and 46(A)). At these points, there was a slight mixing when the elevation reached 1.6m in Piaçabuçu/AL and 1.4m in Ilha das Flores/SE.

The salinity concentration at Points 4, 5 and 6 (Figures 47(A), 48(A) and 49(A)) did not exceed 0.30‰ , with only a stratification of salinity being observed. The salinity concentration of P4, P5 and P6 is classified as fresh water, according to CONAMA resolution 357/2005. Veline (1958) also classifies it as a limnic zone (freshwater) because the salinity values are <0.50‰. This limit is based on the international convention for classifying coastal environments according to salinity due to the ecological importance of these environments. The limit values for classifying the aquatic environment according to salinity are shown in Table 4.

Table 4: Classification of salinity in the aquatic environment

Venice System (1958)	
Zone	Salinity

Hyperhaline	>40%
Euhalina	30% a 40%
Polyhaline	18% a 30%
Mesohaline	5% a 18%
Oligahaline	0,5% a 5%
Limnica	< 0,5%
CONAMA Resolution 357/2005	
Fresh water	Waters with salinity equal to or less than 0.5%
Brackish water	Waters with salinity greater than 0.5% and less than 30%
Salt water	Waters with salinity greater than 30%.

Data on the evolution of salinity dispersion at the mouth of the São Francisco River has been recorded by a number of research studies, including Medeiros (2003), who recorded a salinity of 2‰ in the bottom water 10 km from the mouth, and Medeiros et al (2007), who classify the mixing zone located 8 km from the mouth as mesohaline with a salinity between 5‰ and 20‰. However, Medeiros et al (2014) consider that the environment 6km from the mouth is already home to the salt wedge. This is borne out by the data obtained in this study using salinity simulations.

In general, the points showed an increase in salinity from upstream towards the mouth, with this growth being evident from the municipality of Piaçabuçu/AL, the salt wedge does not remain stationary and seeks a position of equilibrium in response to river and sea discharge. According to Medeiros et al (2001) in a study of the estuary of the River San Francisco, the distribution of salinity is regulated by the flow of the river, the tide and the currents. In his data, the salt wedge was more intense during the period of lowest flow, and at higher flows the salt wedge was displaced towards the mouth. For Medeiros et al (2008), the mouth of the River San Francisco can be classified as having a salt wedge.

According to Pritchard (1952); Hansen and Rattray (1966); Dyer (1997) and Miranda; Castro; Kjerfve (2002) estuaries can be classified according to salinity and circulation patterns. Among these classifications, the types can be saline wedge, partially stratified and mixed.

Salt wedge estuaries are dominated by river discharge and by the process of strangulation, which is responsible for the increase in salinity in the surface layer. Fresh water is less dense than ocean water and tends to remain on the surface, while ocean water has a higher density and penetrates the estuary in the deep layer.

For the partially stratified type, salinity increases gradually, either vertically or horizontally, and the difference between bottom and surface salinity is small. The well-mixed type is classified when the tidal force is dominant over the fluvial force and there is no difference between bottom and surface salinity. According to Segundo (2001), the estuary of the São Francisco River is dominated by the influence of the river and the tide and its classification is partially stratified and well mixed.

The data also infer stratification points in the municipality of Ilha das Flores, 16.21km from the mouth, up to Proprié/SE, 57km from the mouth. It is estimated that the salinity extends up to 15km from the mouth, using the salinity simulation presented in this paper.

Figura 44: Simulated salinity under dry season river flow conditions (A) and water level height with salinity (B) at Point 1.

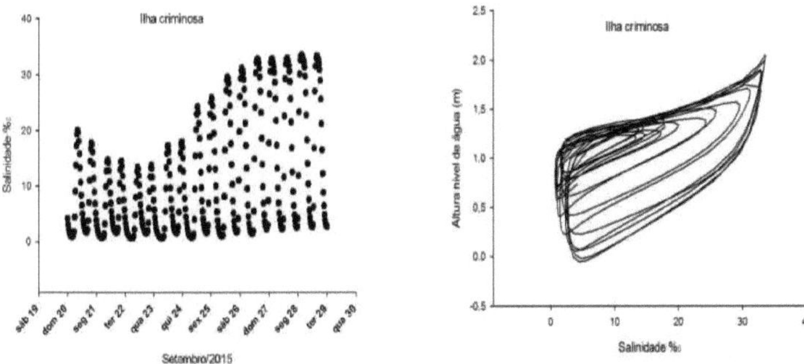

Figura 45: Simulated salinity under conditions of river flow during the dry season (A) and water level height with salinity (B) at P2.

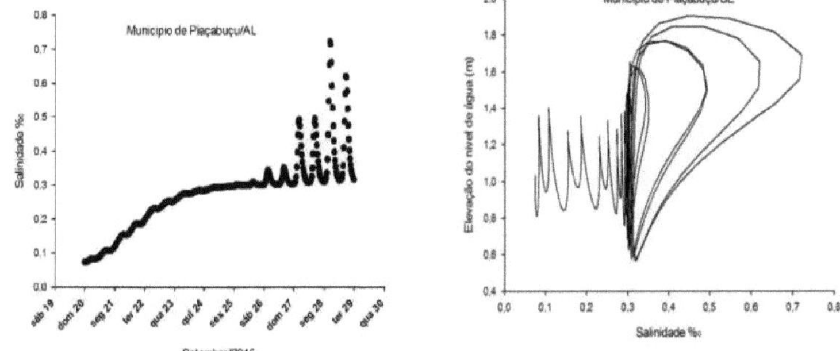

Figura 46: Simulated salinity under dry season river flow conditions (A) and water level height with salinity (B) at P3.

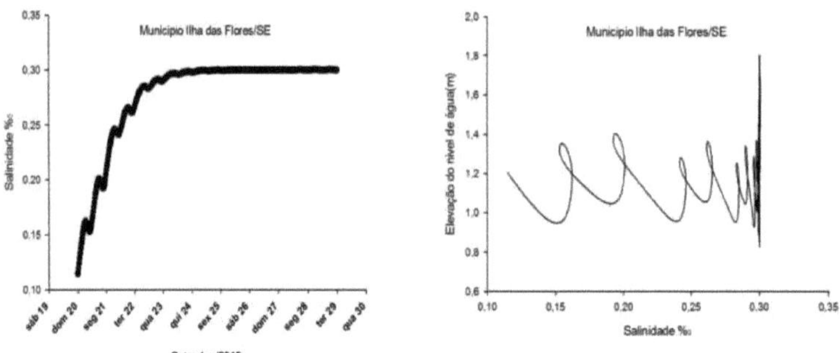

Figura 47: Simulated salinity under dry season river flow conditions (A) and water level height with salinity (B) at P4.

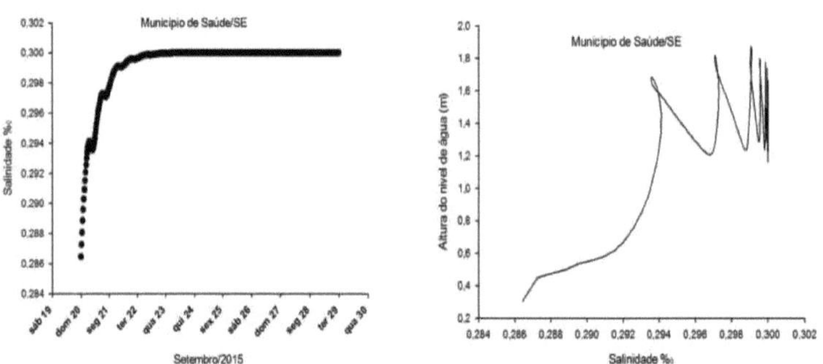

Figura 48: Simulated salinity under dry season river flow conditions (A) and water level height with salinity (B) at P6.

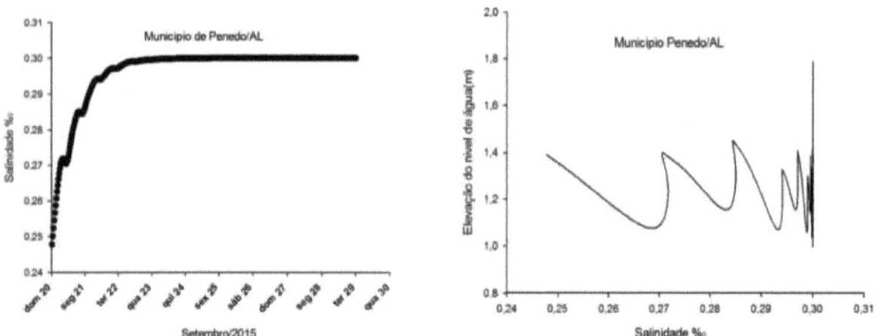

Figura 49: Simulated salinity under dry season river flow conditions (A) and water level height with salinity (B) at P5.

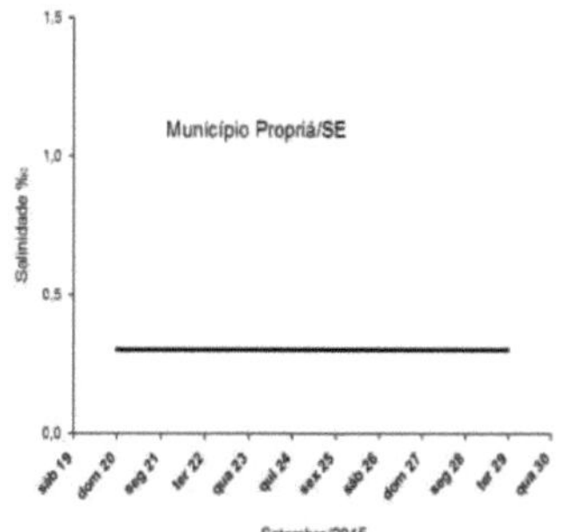
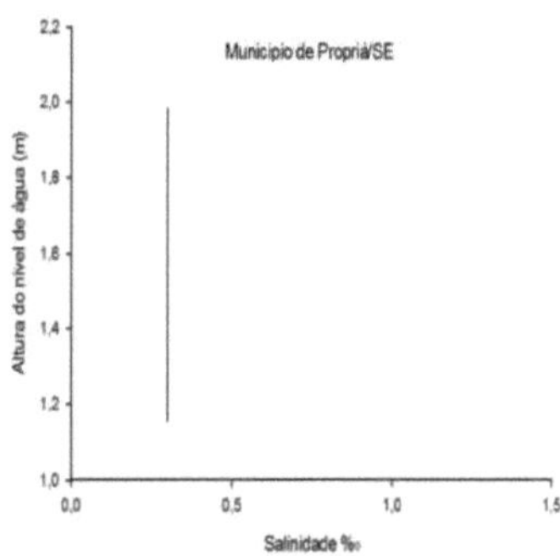

4.3.3 Salinity scenarios at the mouth of the São Francisco River: flood tide and ebb tide.

The sizigia tides are those that represent the highest tidal elevations, so the conditions shown in Figures 49 and 50 record the assessment of the penetration of the salt wedge at the mouth of the São Francisco River and its concentrations identified by color.

Figure 50 shows the result of the salinity simulation by SisBAHIA for September 26 at 2:30 p.m., when the tide rose 2.2m, reaching high tide and the flood situation. Salinity at this time was 32.79‰. Cavalcante, Miranda and Medeiros (2017), evaluating the salt balance in the estuary of the River San Francisco, recorded values of 36.6‰ and 36.8‰ at the mouth in spring at high tide, similar to the data presented in this study.

The result of the salinity simulation using SisBAHIA for 09/27 at 03:30h with a tide elevation of 2.3m reaching the high tide and the flood situation, the salinity concentration for that instant of time was 32.14‰ . In Figures 50 and 51, the rock coloration represents the first moments of the salt wedge's displacement, starting on the left bank of the estuary located in the state of Alagoas.

According to Fontes (2016), in the coastal zone between Alagoas and Sergipe, the maximum tidal amplitudes occur in March and September. According to Castro and Huber (2012) the salt wedge moves with the rhythm of the tides, and in estuaries it moves northwards at high tide and retreats southwards. This means that human individuals and aquatic flora and fauna in this ecosystem are subject to drastic salinity fluctuations.

The results of the simulation for 26/09 showed that in sizigia conditions, a period characteristic of floods, and during the dry season, the saline wedge is distributed from the beginning of the mouth to Ilha Negra, approximately 8km away. It should be noted that at this point the simulated salinity was 2‰, above Piaçabuçu the simulation showed salinity of around 0.3‰, Figure 51. However, on 27/09, the saline intrusion extends as far as Piaçabuçu/AL.

Another aspect presented in the simulation is the distribution of salinity between the tributaries on the left bank located in Brejo Grande/SE, such as the Paraúna river and the Parapuca canal. It should be noted that in the Potengi River the simulated salinity for 26/09 and 27/09 was around

23‰. It should be noted that at this point water is collected to supply the village of Saramém, which is why the community is unable to use the water for consumption at high tide. Record of the collection pumps for the population of Saramém and Brejo Grande/SE (APPENDIX).

For the Potengi River located on the right bank of the mouth, a tributary located in the municipality of Piaçabuçu/AL, the salinity ranged from 7‰ to 24‰. There is no water collection pump in this channel, but the local community takes drinking water directly from the river, and at high tide they usually dig wells in the sand dunes near the mouth in search of fresh water to use (APPENDIX).

In the ebb tide period, the simulation showed that even with the retreat of the tide there is still a permanence of salinization in the Paraùna rivers and in the Parapuca channel at around 9‰ and 15‰. According to Fontes (2016), the formation of this channel was promoted by the trapping of an island colonized by an extensive mangrove swamp in an environment influenced daily by the tide. This justifies the trapping of salts at low tide. In the Paraùna River, 3‰ was recorded on 26/09 and 8‰ on 27/09.

It can be seen that at low tide, the saline wedge is located on the left bank in the municipality of Brejo Grande/SE, close to the Parapuca channel, and was more extensive on 27/09, but there was also an occurrence of salinity trapping on 26/09.

Salinity simulations using SisBAHIA in tidal areas were carried out by Amaral (2003) in the estuary of the Macaé River/RJ, recording salinity of 3‰; Rigo (2004) in the Bay of Vitória/ES with recorded values of up to 35‰; Santos (2012) in the mouth of the Araguaia River/AP with records of 4.83‰; Couto (2014) in the lower course of the Paraguaçu River/BA with salinity records of 5‰ to 35‰; Roversi, Rosman and Harari (2016) in the Santos estuary system, with salinity distribution between 4‰ and 32‰, the highest incidences between 4‰ and 8‰.

Figura 50:   Salinity simulation during flood tides between Ilha criminosa/SE and the municipality of Propria/SE 26.09.

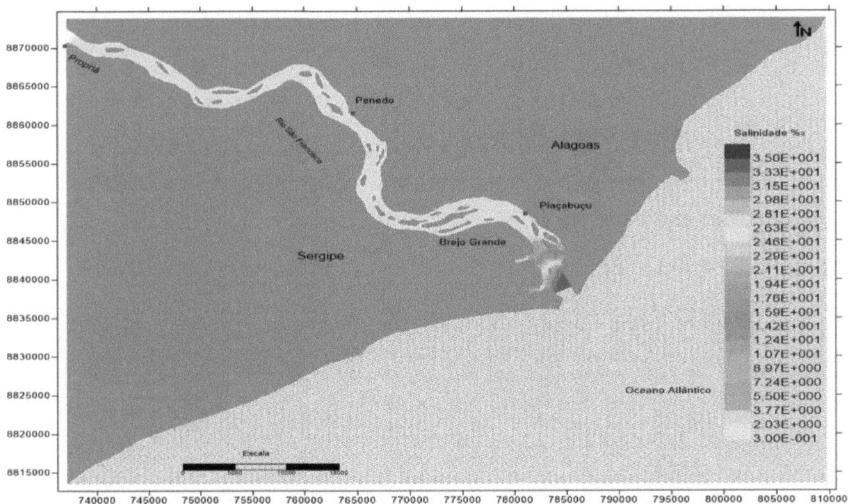

Figura 51: Salinity simulation during high tide between Ilha criminosa/SE and the municipality of Propria/SE 27.09.

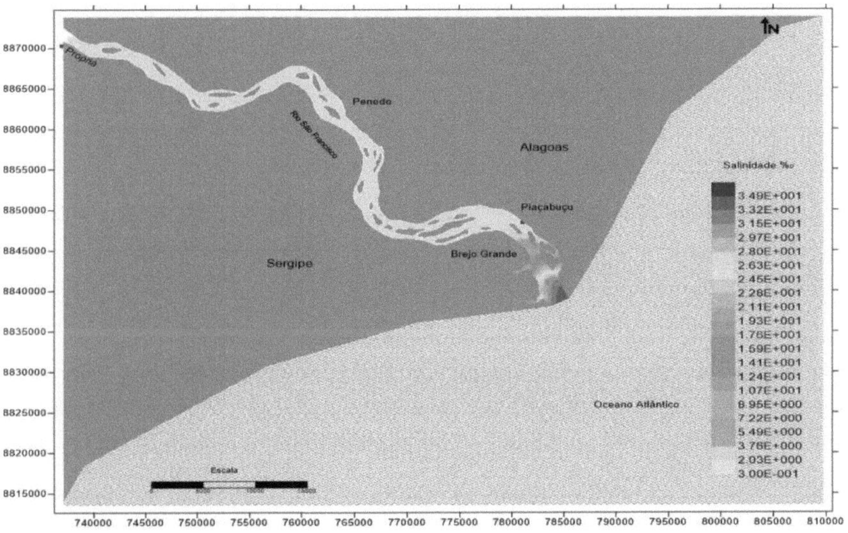

4.4 Conclusion

In conclusion, the SisBAHIA model represented the salt transport mechanism well along the study area. The data showed that salt transport is governed by the action of the tide and has a stratified distribution along the mouth of the River San Francisco, as proven by the calibration and validation of the data.

Salinity is more concentrated in the flood tide regime and in the ebb tide period. The Paraùna and

Potengi rivers showed a brackish water profile and the Parapuca channel showed a saline current.

It should be emphasized that simulating the distribution of salinity can signal the aquatic ecosystem's response to the impacts caused by damming the river, including the limitation of uses for human supply, agriculture and fishing.

It was noted that the salt wedge remains at flood tide approximately 11 km from the mouth between the municipalities of Brejo Grande/SE and Piaçabuçu/AL.

REFERENCES

AMARAL, K. J. Macaé River Estuary: Computer Modeling as a Tool for Integrated Water Resources Management. Thesis. Federal University of Rio de Janeiro/COPPE. 2003. 169p.

AGUIAR NETTO, A. O; LUCAS, A. A. T; SANTOS, A. G. C; ALMEIDA, C. A. P. **Agua e ambiente no baixo Sao Francisco sergipano**. In: AGUIAR NETTO, A. O; LUCAS, A. A. T. Aguas do Sao Francisco. EDUFS. 2011. 15-32p.

CASTRO, P.; HUBER, M. E. **Marine biology**. 8ª ed. AMGH. 2012. 436p.

CAVALCANTE, G. MIRANDA, L. B.; MEDEIROS, P. R. P. Circulation and salt balance in the Sao Francisco river Estuary (NE/Brazil). Revista Brasileira de Recursos Hidricos Brazilian Journal of Water Resources. On-line version ISSN 2318-0331 RBRH, PortoAlegre, v. 22, e31, 2017.

D'AQUINO, C.A; FRANKLIN DA SILVA, L.; COUCEIRO, M.A.A; PEREIRA, M.D. Salt Transport and Hydrodynamics of the Tubarao River Estuary - SC, Brazil RBRH. Revista Brasileira de Recursos Hidricos Volume 16 n.3 - Jul/Sept 2011, 113-125

CBHSF - São Francisco River Basin Committee. Water Resources Plan for the São Francisco River Basin. Monitoring and Evaluation of the Introduction of the Saline Wedge into the San Francisco Estuary. 2009. Available at:<

http://licenciamento.ibama.gov.br/Recursos%20Hidricos/Integracao%20Sao%2
0Francisco/Relatorios%20execucao%20PBA/REL%20SEMESTRAL%207/Prog
rama%2036/Anexo%204.36.1 %20Cunha%20Salina%20Jan-Jul09.pdf> Accessed on September 20, 2016.

DYER, K. R., Sediment transport processes in Estuaries: in: Geomorphology and Segimentology of Estuaries. G.M.E. Perillo (Ed) development in Sedimentology, 53 ElsevierScience. 1995 p.423-449

CASTRO, P.; HUBER, M. E. Marine biology. 8ª ed. AMGH. 2012. 436p.

COUTO, G. A. Análise da influência do regime de vazao da uhe de pedra do cavalo no comportamento espacial e temporal da salinidade no stretch fluvio estuarino do baixo curso do rio Paraguaçu à baia do Iguape/BA. Master's degree dissertation - Federal University of Bahia. Polytechnic School, 2014.143p

FONTES, L. C. SILVEIRA. FROM SOURCE TO BASIN: CONTINENT-OCEAN INTERACTION IN THE SEDIMENTARY SYSTEM RIO SÂO FRANCISCO, BRAZIL. Thesis (PhD) - Paulista State University, Rio Claro Biosciences Institute. 2016, 315p

HANSEN D. V.; RATTRAY M. 1966. New dimensions on estuarine classification. LimnologyandOceanography,11: 319-326.

IPEA. Institute for Applied Economic Research. Immigration project in the lower Sao Francisco Valley. Brasilia: Ipea, 1992, 44p. Available at http√repositorio. ipea.gov.br/bitstream/11058/2517/1/TD%20268.pdf Accessed on March 1, 2016.

LERCZAK, J. A.; GEYER, W. R.; CHANT, R. J. Mechanisms driving the time dependent salt flux in a partially stratified estuary. Journal of Physical Oceanography, v. 36, n. 12, p. 2296-2311, 2006. http√dx.doi.org/10.1175/JPO2959.1.

LEWIS, R. E.; LEWIS, J. O. The principal factors contributing to the flux of salt in a Narrow, partially Stratified Estuary. Estuarine and Coastal Marine Science, V. 16, n. 6, p. 599-626, 1983. http√dx.doi.org/10.1016/0272-7714(83)90074-4.

LOITZENBAUER, E. MENDES, C. A. B. Salinity dynamics as a tool for the integrated management of water resources in the coastal zone: an application to the Brazilian reality Revista da Gestao Costeira Integrada 11(2):233-245 (2011).

MEDEIROS, P.R..P.; SANTOS, M. M.; CAVALCANTE, G. H.; SOUZA, W. F. L.; Environmental characteristics of the Lower Sao Francisco River (AL/SE): effects of dams on the transport of materials at the continent-ocean interface. Geochimica Brasiliensis v 28, n1, 2014. p 65-78.

MEDEIROS P.R.P., Knoppers B.A., dos Santos Jr R.C., Souza W.F.L. 2007. Fluvial input and dispersion of suspended particulate matter in the coastal zone of the São Francisco River (SE/AL). Geoch Bras, 2:209-228.

MEDEIROS P.R.P., Knoppers B.A., Souze W.F.L., Oliveira E.N. 2011. Transport of suspended material in the lower Sao Francisco River (SE/AL) under different hydrological conditions. Braz J Aquat Sci Technol, 15:42-53

MEDEIROS, P.R..P.; SANTOS, M. M.; CAVALCANTE, G. H.; SOUZA, W. F. L.; Environmental characteristics of the Lower Sao Francisco River (AL/SE): effects of dams on the transport of materials at the continent-ocean interface. Geochimica Brasiliensis v 28, n1, 2014. p 65-78.

MIRANDA, L.B.; CASTRO, B.M.; KJERFVE, B.; Principios de oceanografia fisica de estuarios. EDUSP - Editora da Universidade de Sao Paulo. Sao Paulo, Brazil. 2002. 424 p. ISBN: 85-314-0675-7

MONTEIRO, S. M; EL-ROBRINI, M; CASTRO ALVES, I. C. Seasonal dynamics of nutrients in an Amazon estuary. Mercator, Fortaleza, v. 14, n. 1, p. 151162, jan./abr. 2015.

PEREIRA, Marçal D.; SIEGLE, Eduardo; MIRANDA, Luiz B. de and SCHETTINI, Carlos A.F..Hydrodynamics and seasonal suspended particulate matter transport in a sea dominated estuary: Caravelas Estuary (BA). *Rev. Bras. Geof.* [online]. 2010, vol.28, n.3, pp.427-444.

PRITCHARD, D. W.The dynamic structure of a coastal plain estuary. J. Mar. Res. 1952. v15, p 33-42

RIGO, D. Flow analysis in estuarine regions with mangroves - measurements and modeling in the Vit0ria/ES bay. Thesis. Federal University of Rio de Janeiro/COPPE. 2004. 170p.

ROSMAN, P.C.C. A computational system for environmental hydrodynamics. In: Silva, R.C.V. (Ed.), Numerical Methods in Water Resources - Volume 5. ABRH. p. 1-161.2001.

ROSMAN, P. C. C., (Ed.). SisBaHiA® technical reference. COPPE - Federal University of Rio de Janeiro, March 2015, Available at: <http://www.sisbahia.coppe.ufrj.br/SisBAHIA_RefTec_V95.pdf>. Accessed on: June 2016.

ROSMAN, P. C. C., CUNHA, C. L. N., CABRAL, M. M. et al. SisBaHiA Technical Reference. COPPE /UFRJ. Ocean Engineering Program, Rio de Janeiro, RJ. Version 9.2, 2013. 249 p. Available at:

http://www.sisbahia.coppe.ufrj.br/SisBAHIA_RefTec_V92.pdf. Accessed on: September 20, 2015.

ROVERSI, F.; ROSMAN, P. C.; HARARI, J. Analysis of the Trajectories of Continental Waters Flowing into the Santos Estuarine System. RBRH vol. 21 no.1 Porto Alegre jan./mar. 2016 p. 242 - 250.

SANTOS, E. S. Hydrodynamic modeling and water quality in a pororoca region at the mouth of the Araguari River/AP. Dissertation. Federal University of Amapa Foundation. 2012. 113p.

SEGUNDO, G. H. C. Sedimentological hydrodynamic characterization of the estuary and delta of the São Francisco River. Dissertation meteorology department/CCEN/UFAL. 2001. 102p

VENICE SYSTEM. Symposium on the classification of brackish Waters, Venice, April 8-14, 1958. **Archives of Oceanography and Limnology**, v. 11, p. 1-248, 1958.

CHAPTER 5. ENVIRONMENTAL EDUCATION ACTIVITIES IN THE REGION OF THE MOUTH OF THE RIVER SAN FRANCISCO: SOCIOECONOMIC PROFILE OF FISHERS AND ENVIRONMENTAL DEGRADATION

SUMMARY

Over the years, the relationship between man and nature has caused a scale of intervention in the dynamic balance of the environment. In recent decades, the São Francisco River Basin has undergone a period of dam construction, which has been responsible for major natural and social changes. This study was carried out in the area of the mouth of the River San Francisco, which covers a stretch of coastline about 25 km to the south, up to the town of Ponta dos Mangues, in the municipality of Pacatuba. The tool used to analyze perceptions was based on the application of a semi-structured questionnaire with the aim of collecting information on the socio-economic profile of fishermen, knowledge of the type of aquatic plants and other factors that harmed the river. From the data obtained, the predominant age profile of those interviewed was over 45 years old, with a significant female presence. The perceived degradation pointed to aquatic plants as being responsible for the loss of equipment and the reduction of fish, reports of the migration of saltwater fish species in the river is already common, and the disposal of solid waste was also characterized as one of the elements of degradation in the region. It is concluded that the environmental education activities enabled fishermen to understand the main types of environmental degradation in the region investigated.

Key words: Fishermen, Environmental degradation, Environmental perception.

5.1. Introduction

Over the years, the relationship between man and nature has led to a scale of intervention in the dynamic balance of the environment. The product of this relationship is human activity, which has had an impact on the environment and altered natural characteristics.

In recent decades, the São Francisco River Basin has undergone a period of dam construction, which has been responsible for major natural and social changes in the area, with negative environmental impacts that were not considered when they were installed.

This modification along the river favors numerous imbalances that can affect the fauna, flora and abiotic processes of an aquatic environment. These changes are caused by alterations to aquatic species, such as possible types of macrophytes.

According to Araújo et al (2012), aquatic macrophytes have different morphophysiological adaptations, as well as the ability to colonize aquatic environments with different physical and chemical characteristics, and have a wider distribution than most terrestrial plants, a distribution that occurs due to variations in the aquatic environment.

Another aspect that signals imbalance is the disappearance of freshwater fish species in the area of the mouth of the River San Francisco and the migration of new species to adapt to the modified spaces. In addition to these factors, the disposal of solid waste and the removal of the forest on the banks have been the subject of dialogues that demand the River San Francisco through activities in the area of environmental education, since it is understood that through knowledge there is the necessary condition to modify a picture of growing socio-environmental degradation in the region.

Environmental education must be introduced with caution, taking care to cover the environmental crisis on different scales and relating it to the existing dysfunctions in the style of development, always starting from the principle that the social issue and the environmental issue are intrinsic to each other (LONGO, 2016).

In view of the above, the aim of this study was to verify a perceptual diagnosis of environmental degradation identified by fishermen in the area around the mouth of the River San Francisco, seeking to understand the main causes and consequences of the imbalance occurring in the region.

5.2. Material and methods

5.2.1 Study area

The São Francisco River rises in the Serra da Canastra mountains in Minas Gerais and flows into the Atlantic Ocean on the border with the states of Alagoas and Sergipe. In the case of the São Francisco, its mouth is an estuary, as it is made up of a long channel with a tapering shape and, as the river approaches the sea, its banks become closer together (Figure 52) (HERMUCHE, 2002).

The area of direct influence of the estuary covers a stretch of coastline about 25 km to the south, up to the town of Ponta dos Mangues, in the municipality of Pacatuba. To the north, there is approximately 18 km of beach to the town of Pontal do Peba (AL) (Valente et al, 2011). The mouth of the São Francisco River is one of the main natural beauty spots in the lower São Francisco region and has a major influence on the fishermen's way of life.

Figure 52: Geographical location of the São Francisco river mouth area

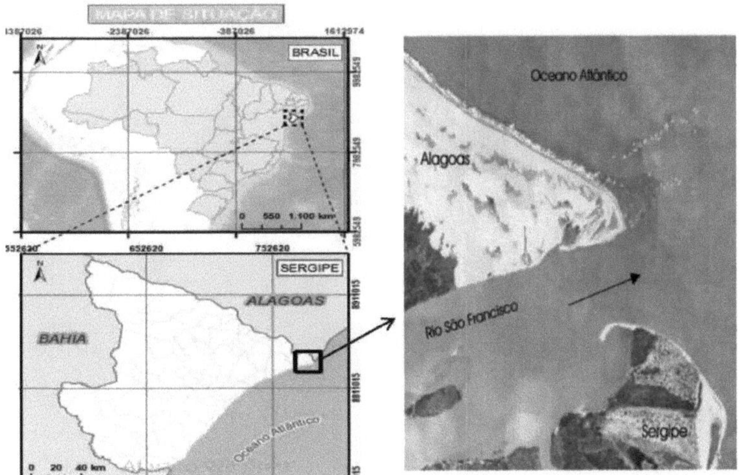

Source: IBGE: Political-Administrative Division of Brazil. 2002. SEMARH-SE: Division

Source: O rio de Sâo Francisco

(HERMUCHE, 2002).

Municipal; Hydrography; State Highway; Cities. 2012 .Elaboration: Luciano Lima .Universal Transverse Mercator Projection

5.2.2 Data collection _ ....

To begin collecting data, the head of the Fishermen's Association in the area of the mouth of the River San Francisco was asked to present the research project with themes related to environmental degradation in the River San Francisco. The tool used to analyze perceptions was based on the application of a semi-structured questionnaire (APPENDIX C) from May to June 2015 in order to collect information on the socio-economic profile of the fishermen, knowledge of the type of aquatic plants that harm the river, types of degradation, and whether there were conservation actions in the river area.

To complement the interaction of the information, an album was created with photographic images of the possible aquatic macrophytes present in the area of the mouth of the River San Francisco (Figure 53A). The dynamics of this activity took place in groups where the interviewees looked at the images and then wrote down their answers and gave their personal opinions on the subject under investigation (Figure 53B). There were non-literate interviewees in the group, who sought help in transcribing their information, but without ideological intervention. All data collection consisted of a qualitative and quantitative approach, according to Richardson (1999), where the quantitative method represents, in principle, the intention to guarantee the accuracy of the results and the qualitative method is justified because it is the appropriate way to understand the nature of a social phenomenon.

Figure 53: Images of aquatic macrophytes used as a visual resource during interviews with fishermen (A) and application (B).

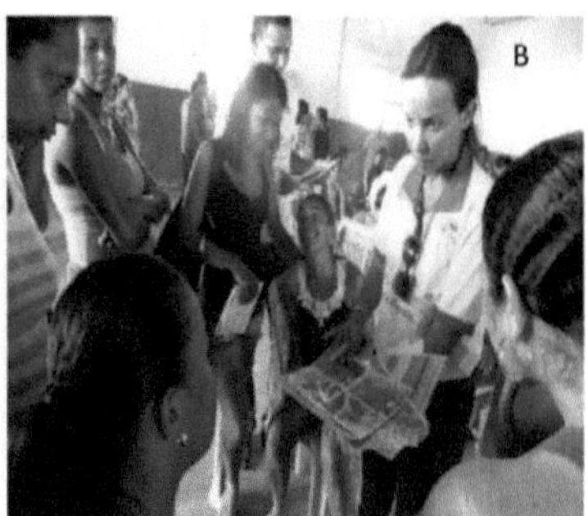

Source: Personal collection (2015)

5. 2. 3 Data analysis

The results obtained from the objective questions were tabulated and coded using spreadsheets, after which the data was treated statistically by mean and frequency. The other observations were interpreted in order to synthesize the data on the object of the research (RAMPAZZO, 2011).

5.3. Results and discussion

5. 3.1 Fishermen's profile and perception of environmental degradation in the Foz do Rio Sao Francisco area

The results obtained allow us to infer that the predominant age profile of the interviewees is over 45 years old (Figure 54A), similar to the values found by Ramires (2012) in his research in the Ribeira Valley and Pontal de Sao Paulo. At the time of this research, there was a predominance of women, and from observation, most of them take on or contribute to the family income (Figure 55A). Garcez and Botero (2005) point out that women are already considerably active in fishing, and are even professionally documented. Alencar and Maia (2011) add that national data shows that women are becoming more involved in fishing, especially in the north and northeast of the country.

The level of schooling of the population evaluated is concentrated between elementary school and incomplete high school (Figure 54B). This reality was also detected by Tamano et al (2015), whose sample group indicated an education level of between 5^a and 8^a grades. It is noteworthy that some of the interviewees had higher education, a fact that may be related to government actions in recent years, which enable individuals who live far from urban centers to have access to professional training. When it comes to assessing income, according to the information declared in the interviews, the main source of livelihood is based on fishing and the family grant (a resource provided by the federal government), see Figure 55A, but almost 22% of fishermen over the age of 40 received a pension.

Regarding the degradation of the river, the main factors pointed out by the interviewees were the disposal of solid waste in the river itself and dams (Table 16 and Figure 54B). They emphasized that damming reduces the amount of water in the river, making fishing and navigation difficult, as well as promoting the formation of sandbanks along the course of the river. Regarding the presence of garbage, they stressed that it is often disposed of by the fishermen themselves and that they feel there is a need for educational and enforcement action to reduce this practice. With regard to the conservation of the River San Francisco, 70.60% of those interviewed said no and 27.5% said yes (Table 17).

Figura 54: Age group (A) and level of education of those interviewed in the municipality of Brejo GrandeZSE (B).

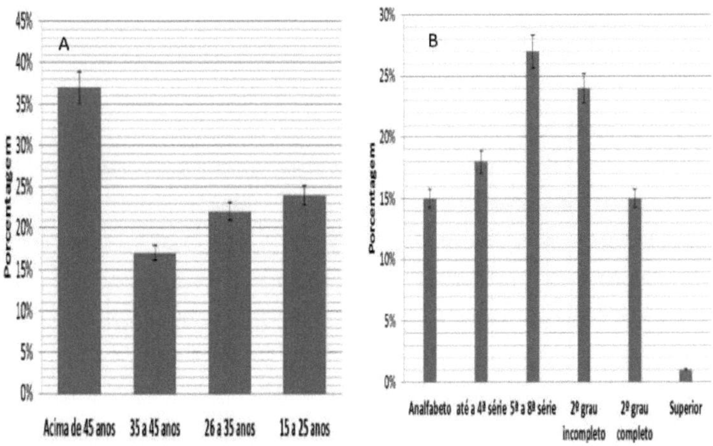

Figura 55:   Income activity (A) and causes of degradation in the Foz do Sao Francisco area (B).

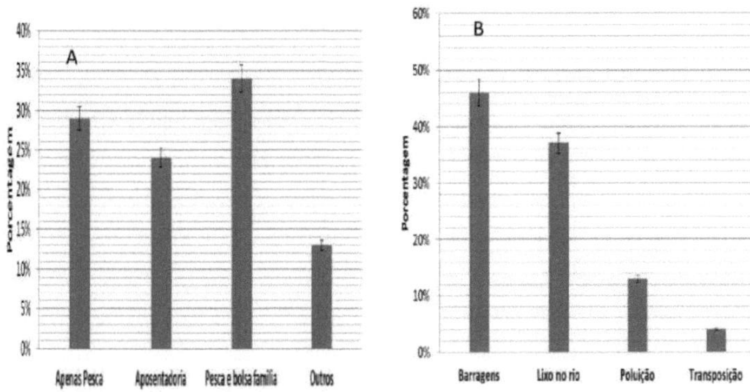

Source: Personal collection

Table 16: Fishermen's perception of degradation in the area of the mouth of the River San Francisco.

Reasons for the degradation of the San Francisco River

Answers	Frequency	Percentage
Uespejo de Kesiduos	5ŏ	52,34
Dam influence	19	17,58
General pollution	19	1/,58
sanitary sewage	13	12,15

Source: Author (2017)

Table 17: Fishermen's perception of the conservation of the River San Francisco.

Conservation of the River Sâo Francisco

Answers	Frequency	Percentage
No	113	/0,6
Yes	44	2/,5
Doesn't know how to form	3	1,9

Source: Author (2017)

Based on the data obtained during the survey, it was found that most of the interviewees had a personal interest in giving a detailed account of the main difficulties encountered as a result of the increase in macrophytes in the course of the River San Francisco. All of them were involved in fishing and sailing, with women accounting for almost 67%.

With the formation of ten focus groups, they began to sort the images of the macrophytes and select the question "Which of these macrophytes hindered fishing and navigation". By writing and speaking, the groups reported and indicated the species by the popular name known in the region. Among the aquatic macrophytes mentioned were *Egeria densa,* popularly known as "Cabelo", and *Eichhornia crassipes,* popularly known as Baceiro (Figure 56). In addition to these, *Derbesia tenuissima,* commonly known as "slime", was also classified; Figure 56 shows the representation of the grouping of information.

The main complaint of the interviewees was that these vegetations restrict fishing yields, due to the entanglement in the net or in the propellers of the boats, causing them to interrupt their activity to remove these species from their equipment. One of the interviewees also said "*Baceiro for those who have a boat and Cabelo for those who have a net*", referring to the fact that the Baceiro vegetation was the most damaging for navigation, due to its agglomeration in the river *"The more the river gets shallow it forms croa, where it was deep it gets shallow", as* well as damaging the boat's propeller, reducing navigation in these areas.

This statement refers to the decrease in water flow and erosion processes along the lower São Francisco, making it possible for the water level to decrease and sandbanks to form. Santana, Araújo and Vasco (2015) point out that navigation activity is already hampered in several stretches due to the accumulation of sediment, reinforcing the discourse presented by the interviewee.

The sand banks formed by the accumulation of sediment not only hinder navigation, but also become a favorable place for macrophytes to settle, favoring their proliferation and limiting socio-environmental activities in these localities. Silva, Marques and Lolis (2012) emphasize that the formation of extensive banks of aquatic macrophytes can hinder people's access to water resources. However, the interviewee still says that this is because there are no more floods, "*the water is trapped and it only grows*".

As for the vegetation known as Cabelo, he said that when he throws the net it doesn't go down to greater depths because the vegetation forms barriers and clings to the structures, preventing the fish from entering. He also said that *"When they release the water, it already cleans the Cabelos"*, referring to the reduction in the flow of the São Francisco river, which today stands at 800 m^3/s according to CHESF (Companhia Hidro Elétrica do São Francisco)/2016. The interviewee finishes his statement by saying, "*It's coming from the middle to the edges, harming fishermen.* Emphasizing that the aquatic macrophytes circulate to the main course of the river, harming fishing activity.

The species *Montrichardia arborescens*, popularly known as Aninga, was highly praised by the interviewees, who said that this vegetation is important for maintaining the river, as it prevents siltation and is also home to fish and crustaceans. This highlights the fact that siltation is one of the most widespread impacts on the banks of the mouth of the River San Francisco. For Nascimento, Ribeiro Jr and Aguiar Netto (2013), after the regularization of flows, coastal dynamics changed substantially and erosion became dominant to the south and north of the mouth. This scenario may explain the increase in erosion in the region, but it should be noted that the loss of vegetation in the marginal zones also contributes to the fragility of the soil and increases erosion along the river.

For Pompeo (2008) it has already been proven that aquatic macrophytes proliferate in environments with high nitrogen and phosphorus concentrations, but extensive banks of floating macrophytes have been observed in places with low nitrogen and phosphorus concentrations.

Other correlations to understand the growth of this vegetation may be related to climate, temperature, the construction of artificial reservoirs or anthropogenic pressure (CAMARGO, PEZZOTO and SILVA, 2003).

Bianchini Junior (1998) points out that the development of *Eichhornia crassipes* has intensified in tropical regions, as it tends to occupy the surfaces of canals, rivers, lakes and reservoirs (Figure 57A). These species end up causing problems in environments that have been altered by humans, and one of the propitious environments for the spread of these species is reservoirs for the purpose of generating energy, representing a major economic problem for the country. It is also worth mentioning Pitelli's (1998) records on the reduction of herbivorous fish in the watercourse which allows a large population flow of *Egeria densa* (Figure 57B).

Figure 56: Profile of the interviewees' responses on the species of macrophytes that hinder navigation.

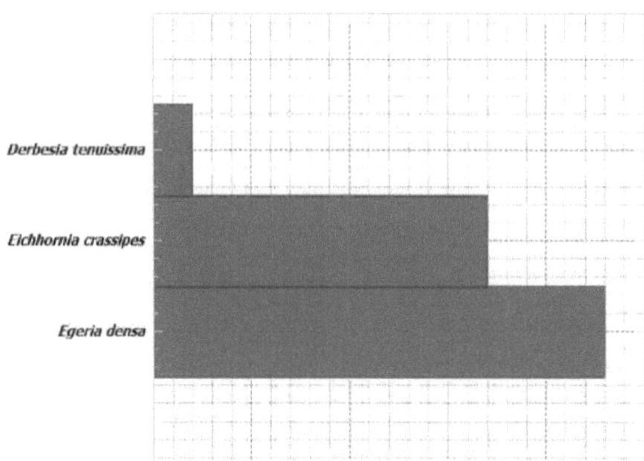

Figure 57: Images of the aquatic macrophytes *Eichhornia crassipes (A) and Egeria densa* (B) in the Foz do Rio Sâo Francisco area, municipality of Brejo Grande/SE.

Another important point raised by the fishermen is the absence of freshwater fish species that were once common in the region and are now considered to have disappeared. Fish species such as the Surubim *(Pseudoplatystoma corruscans)* and Xira (*Prochilodeu sargenteus*) are no longer common in the lower Sâo Francisco. According to Costa et al (2003), these species are already considered to be under threat of extinction. The author emphasizes that the Surubim *(Pseudoplatystoma corruscans)* is one of the most important fish species for the Sâo Francisco River in biological and fishing terms.

The fishermen also pointed out the presence of saltwater fish in the river, such as stingrays *(Dasyatis marmorata)* (Figure 58). According to them, as the river has lost its strength due to the decrease in the flow of water from the headwaters to the mouth, saltwater fish species are occupying habitats that used to belong only to freshwater fish species. When asked why they thought the river was getting saltier, interviewee 2 described it as *"the river is drying up and the sea is invading", while interviewee 40* said *"the river is drying up"*. These two statements were confirmed by almost 70% of the fishermen.

Figure 58: Catch of ray among the fish caught in the Foz do Rio Sâo Francisco area.

Source: Personal collection

At the end of the interviews, the aim was to present a lecture on the importance of the River San Francisco and the need to conserve this resource (Figure 59). This was a moment of integrating *on-site* and scientific knowledge and really bringing to the fore environmental problems in their interrelationship. In this light, Silva and Magalhâes (2014) state that the emergence of discussions and reflections on this problem signals a change in behavior as fundamental to the internalization of sustainable habits.

Figure 59: Environmental education activity at the fishermen's colony in the municipality of Brejo Grande/SE.

5.4 Conclusion

In conclusion, the data obtained from the fishermen's speeches made it possible to see the scale of the main degrading elements in the region of the mouth of the River San Francisco. Interesting factors such as the fact that women account for the majority of the representation during the activity have led to a change in the fishing routine, and that a large part of their source of income comes from fishing, which justifies the real importance of environmental education activities in this location. With regard to the perception of aquatic macrophytes, the fishermen indicated that they were well aware of their consequences for socio-economic activity and related these problems to the dams introduced into the river channel. Although this vegetation occurs naturally in aquatic environments, its upwelling is causing direct disturbances to navigation and fishing activities. It is worth noting that of the 4 species presented, *Eichhornia crassipes* and *Egeria densa* were highlighted as being responsible for causing damage and reducing the efficiency of the equipment used by fishermen during their working day.

REFERENCES

ALENCAR, C. A. G; MAIA, L. P; Socioeconomic profile of Brazilian fishermen. **Arq. Cièn. Mar**, Fortaleza, 44(3): 12 - 19.2011.

ARAUJO, E.S.; SABINO J.H.F.; COTARELLI, V.M.; ALVES, J.A.S.; CAMPELO, M.J.A.; Richness and diversity of aquatic macrophytes in Caatinga springs. **Dialogos & Ciència**, Petrolina, DOI:10.7447/dc.2012.

BIANCHINI JUNIOR, I. Models of growth and decomposition of aquatic macrophytes In: THOMAZ, S. M; BINI, L. M. Ecologia e manejo de macrófitas aquaticas. Maringa: EDUEM, 2003. 342p.

CANCIAN, L. F. Growth of the floating aquatic macrophytes Pistia stratiotes and Salvinia molesta under different temperature and photoperiod conditions. Dissertation (master's degree in Aquaculture) - Universidade Estadual Paulista, Jaboticabal, SP. p. 66. 2007.

COSTA NETO, S. V; SENNA, C. do S. F; TOSTES, L. de C. L; SILVA, S. R. M. Macrófitas aquaticas das Regioes dos Lagos do Amapa, Brasil. Revista Brasileira de Biociencias, V.5, p. 618-620. 2007.

COSTA, F.J.C.B. (coord). Recomposition of the ichthyofauna of the Lower São Francisco River. Project for the Integrated Management of Activities carried out on Land in the São Francisco Basin. **Final report**. Canindé de Sao Francisco: Instituto de Desenvolvimento Cientifico e Tecnològico de Xingó. 2003.

CUNHA, M. B; BIANCHINI JR. I. Colonization of aquatic macrophytes in lentic environments. Rio Claro-SP: Boletim ABLimno, Universidade Estadual Paulista (UNESP), vol. 39(1), 2011.

DINIZ, C. R.; CEBALLOS, B. S. O.; BARBOSA, J. E. de L; KONIG, A. "Use of aquatic macrophytes as an ecological solution for improving water quality". Revista Brasileira de Engenharia Agricola e Ambiental, v. 9, pp. 226230. 2005.

GARCEZ, D. S; BOTERO, J. I. Caracterizapao da pesca artesanal e o conhecimento pesqueiro local no vale do Ribeira e litoral sul de Sao Paulo. **Revista Atlàntica**, Rio Grande do Sul. 17-29. 2005.

GODINHO, H.P.; GODINHO, A.L. Aguas, Peixes E Pescadores Do Sao Francisco Das Minas Gerais. Belo Horizonte: Pucminas, 2003. 18p.

HERMUCHE, P. M. The Sao Francisco River. Companhia de desenvolvimento do vale do Sao Francisco e do Parnaiba. **CBHS.** 58 p, 2002.

LONGO, G. R. Environmental education and values education in teacher training. **Revista eletrônica mestrado em educação ambiental**. v33. n1. Jan/April. 256-268p. 2016.

MEDEIROS, P. R. P; SANTOS, M. M; CAVALCANTE, G. H; SOUZA, W. F. L; SILVA, W. F.

Environmental characteristics of the Lower Sao Francisco River (AL/SE): effects of dams on the transport of materials at the continent-ocean interface. Geochimica Brasiliensis, 28(1):65-78. 2014.

MMA - Ministério do Meio Ambiente, Caderno da Regiao Hidrografica do Sao Francisco Secretaria de Recursos Hidricos. - Brasilia: MMA, 148p, 2006.

MOURA, M. A. M.; FRANCO, D.A.S.; MATALLO, M.B. Manejo integrado de macrófitas aquaticas. Instituto Biològico, Sao Paulo, v.71, n.1, p.77-82, jan./jun. 2009.

NASCIMENTO, M. C; RIBEIRO JÙNIOR. C. E; AGUIAR NETTO, A. O. Technical report of the campaign to evaluate the socio-environmental changes resulting from the regularization of flows in the lower São Francisco River. CBHSF. Maceio, AL, 2013, 175p.

PEZZATO, M. M; SILVA, G. G. H. BIANCHINI JUNIOR, I. Factors limiting the primary production of macrophytes In: THOMAZ, S. M; BINI, L. M. Ecologia e manejo de macrófitas aquaticas. Maringa: EDUEM, 2003. 342p.

PITELLI, R. A. Aquatic macrophytes in Brazil, in problematic condition. In: WORKSHOP CONTROLE DE PLANTAS AQUATICAS, 1998, Brasilia, **Abstracts...** Brasilia: IBAMA - Brazilian Institute for the Environment and Renewable Natural Resources, 1998. p. 12-15.

POMPEO, M. L. M.. Aquatic macrophytes in tropical reservoirs: ecological aspects and proposals for monitoring and management. In: Pompêo, M.L.M. (Ed.) Perspectivas da Limnologia no Brasil. p.105-119. 1999.

POMPEO, M.; Monitoring and Management of Aquatic Macrophytes. Oecol. Bras., Sao Paulo, v.12, n. 3, p. 206-224, 2008.

RAMIRES, M; BARRELLA, W; MARTUCCI, A. E. Caracterizaçao da pesca artesanal e o conhecimento pesqueiro local no vale do Ribeira e litoral Sul de Sao Paulo. **Revista ceciliana**, June. 37-43, 2012.

RAPAZZO, L. **Metodologia cientifica: For undergraduate and postgraduate students.** Editora Loyola. 6ª ed,146p, 2011.

RICHARDSON, R. J. **Pesquisa social: Métodos e técnicas**. Editora Atlas, 3aed. 1999.332p.

SANTANA, N. R.F; ARAÙJO, S. S; VASCO. A. N. Diversos olhares sobre a Foz do rio Sao Francisco. In: Contexto socioambiental das aguas do rio Sao Francisco. Org. AGUIAR NETTO, A. O; SANTANA, N. R.F. Editora UFS, 2015. P 137-149.

SILVA, D. S; MARQUES, E. E; LOLIS, S. F. Aquatic macrophytes: "villages or young ladies"? Interface, Issue 04, May 2012. ISSN 1806-6062.

SILVA, S. L; MAGALHAES, K. M. Environmental perception of aquatic macrophytes and environmental impacts by students in the metropolitan region of Recife/PE. **Revista eletrônica mestrado em educaçao ambiental.** v31. n1. Jan/June. 174-188p.2014.

SOUZA, L. S; NUNES, R. de O. Survey of aquatic macrophytes in the Méquens River. Cacoal-RO: Revista Cientifica Eletronica FACIMEDIT. V.2, p. 211222. 2010.

TAMANO, L. T. O;ARAUJO, D. M. LIMA, B. B. C; SILVAII, F. N. F. S. J. Socioeconomics and health of Mytellafalcata fishermen from Mundaù Lagoon, Maceió-AL. **Bol. Mus**. Emilio Goeldi. Cienc. Hum., Belém, v. 10, n. 3, p. 699710, Sep.-Dec. 2015.

THOMAZ, S.M.; BINI, L.M; Ecology and Management of Aquatic Macrophytes in Reservoirs; Acta Limnologica Brasiliensia, Maringa, v. 10, n. 1, p. 103-116, 1998.

VALENTE, R. M;SILVA, J. M. C; STRAUBE, F. C; NASCIMENTO, J. L. X. Conservation of Neartic migratory birds in Brazil Conservação de Aves. Conservação Internacional - **CI-Brasil**. 2011.

APPENDICES

Table 1: Measured and simulated salinity values at the reference point.

Date and time	Measured	Simulated
26/9/15 10:30h	2,71	0,82
26/9/15 13:00h	30,08	29,34
26/9/15 14:00h	32,53	33,12
26/9/15 14:30h	35,21	33,20
26/9/15 15:30h	36,04	32,26
26/9/15 16:30h	36,1	27,51
26/9/15 17:30h	24,7	18,32
26/9/15 18:30h	16,17	9,14
26/9/15 19:30h	14,07	4,03

Source: Personal collection

APPENDICES

Figura 1: Drinking water collection pump operated by the Sergipe state sanitation company, located on the Paraùna river.

Figura 2: Drinking water collection pump operated by the state sanitation company of the state of Sergipe, located in Brejo Grande/SE.

APPENDICES

FEDERAL UNIVERSITY OF SERGIPE   PRODEMA - UFS

DEAN'S OFFICE FOR POSTGRADUATE STUDIES AND RESEARCH

Graduate Program in Development and the Environment PHD IN DEVELOPMENT AND THE ENVIRONMENT

SEMI-STRUCTURED INTERVIEW QUESTIONNAIRE
FIELD: MUNICIPALITY OF BREJO GRANDE/SE

PERSONAL DATA

Name:

1.   AGE:( )15 to 25 years(   )26 to 35 years( )35 to 45 years ( )over 45

years

2.   SCHOOLING: ( )Illiterate( )up to 4^a serle( )5 to 8^{aa} serle ( ) 2^0 complete high school ( )20 incomplete high school ( )college degree

3.   DO YOU LIVE IN THIS MUNICIPALITY? ( ) YES ( ) NO

4.   THE HOUSE IS OWNED ( ) YES ( ) NO

5.   HOW MANY PEOPLE LIVE WITH YOU:

6.   DO YOU GET YOUR LIVELIHOOD FROM: FAMILY GRANTS ( ) FISHING ( ) FARMING ( ) HANDICRAFTS ( ) MAKING SWEETS AND COCONUT CAKES ( ) OTHERS: _______

7.   IS THERE RUNNING WATER ( ) YES ( ) NO

WHO SUPPLIES? _______________________________

8.   IS THERE A LOT OF WATER IN THE AREA? ( ) YES ( ) NO

9.   WHY IS THERE NO WATER? _______________________________

10.   Paravoce Is the river well cared for? yes( ) no( ) ii)io) Paravoce What is harming the river?

12)   WHERE DO YOU GET YOUR DRINKING WATER FROM?

13)   WHAT COLOR IS THE WATER YOU DRINK? CLEAR ( ) YELLOWISH ( ) DARK ( )

14)   FOR YOU THE TASTE OF WATER IS: BITTER ( ) SALTY ( ) SWEET ( )

15)   HOW SALTY IS THE WATER?

ALWAYS ( ) HIGH TIDE ( ) LOW TIDE ( )

16)   WHEN THE WATER IS SALTY, DO YOU USE IT TO DRINK? YES (   )

NO ( )

17)   DO YOU USE ANY MEANS TO REMOVE SALT FROM THE WATER? YES (   ) IF YES

WHICH? NO ( )

18)   DOES IT HAVE A FOSSA?

() yes-where is the residence? ()no

19)   WHERE DO YOU DISPOSE OF YOUR HOUSEHOLD WASTE?

20)   ) WHAT FISH USED TO BE IN THE RIVER AND ARE NO LONGER THERE TODAY?

21)   20)WHICH SALTWATER FISH HAVE YOU SEEN IN THE RIVER?

22)   DO YOU THINK THE RIVER IS WELL CARED FOR? YES ( ) NO ( ) 22) FOR YOU THE

THAT IS DAMAGING THE RIVER?

23)   WHICH VEGETATION ON THE RIVER HARMS NAVIGATION? 24) WHICH VEGETATION HARMS THE RIVER THE MOST?

25)WHATWHEN THE WATER IS SALTY?ALWAYS( ) HIGH TIDE ( ) LOW TIDE ( )

26)   WHEN THE WATER IS SALTY DO YOU USE IT TO DRINK? YES ( ) NO ( ) 27)WHY DO YOU THINK THE RIVER WATER IS GETTING SALTY?

ANNEX

HAIAR- HNDA(,Λ( ) DE ESTTI DOS DO MAR *Catalogue of Brazilian Maritime Stations*

Station Name :	HEAD - IF		
I Realization :	In Barra do Rio SAo Francisco, in front of the Lighthouse		
Organ. ResponsAveI :	INPH / DHN		
Latitude :	IO* 30,2' S	Ixmgitude :	36* 24,0' W
Period Analyzed :	06/04/81 a 12/05/81	N* of Components :	82
Harmonic Analysis :	Admiral Santos Franco Method		
Classification :	Marò Semidiurna		
Establishment in Porto: (HWFAC)	IV H 21 min	Middle Level (Zo):	99 cm from the NR.
Mean High Water Mark (MHWS) :	176 cm from NR.	Mean Square Wave Height (MHWN) :	134 cm above NR
Mean Low Water Syzygy (MLWS) :	22 cm above the NR.	Mean Square Low Water (MLWN) :	63 cm above NR.

SELECTED HARMONIC CONSTANTS

Components	Semiamplitude (H) cm	Phase (g) degreesf)	Components	Semiamplitude (H)cm	Phase (g) degrees (*)
Sa	■■	-	MU_1	10,1	049
Ssa	-	*	NJ	16.8	074
Mm	3,7	196	NU_2	3,2	077
Mf	-		M,	56.2	102
MTM	3,2	050	L_2	7,6	123
M$f	1Λ	169	T_1	1,2	127
Qi	1,3	021	¾	20,5	129
O_1	9,4	125	K_2	5,6	131
M_1	2,1	200	MOj	3,3	140
Pi	i,0	221	. M_J	1,6	264
K_1	3,2	229	MK_J	2,4	265
Ji	],8	074	MN_4	1.8	115
OOi	2,4	320	M_4	2,3	148
MNSj	1.9	113	SN_4	1,0	051
$2N_i$	2.2	045	MS_4	1.5	158

Level references: NR Located on the 9th step from the bottom to the top of Slo Francisco do Norie.
Obs :
There are no references to other périodes

BNDO code 30800